PLASTID BIOLOGY

Plastids reside in all plant cells, and take on different forms in relation to their cellular function, biochemistry and storage capacity. The modern era of molecular biology and molecular genetics has enabled much to be learnt about how plastids function, and how they relate to their evolutionary past. In this accessible text, Kevin Pyke expertly describes how the plastids are highly complex organelles at the very core of plant cellular function, providing final year undergraduate and graduate students with an overview of plastid biology and recent developments in the field. Topics covered include: a consideration of different plastid types and how they relate to cell function; plastid genomes and how proteins are imported into plastids; photosynthesis and core aspects of plastid biochemistry; plastid signalling and functionality within a cellular context; and plastid genetic manipulation. Supplementary colour images are available online at www.cambridge.org/9780521885010.

KEVIN PYKE has carried out research into various aspects of plastid biology over the past 25 years. He is Associate Professor in Plant Cell Biology in the Plant and Crop Sciences Division of the School of Biosciences, University of Nottingham. He has also worked at the John Innes Institute, Norwich, the University of York, and Royal Holloway, University of London. He identified an important collection of mutants in Arabidopsis in which chloroplast division was perturbed, which led to the identification of several novel genes functional in this process. More recently, he has worked on stromules and how they might enhance plastid function within the cell.

Plastid Biology

KEVIN PYKE
University of Nottingham, UK

CAMBRIDGE
UNIVERSITY PRESS

Shaftesbury Road, Cambridge CB2 8EA, United Kingdom

One Liberty Plaza, 20th Floor, New York, NY 10006, USA

477 Williamstown Road, Port Melbourne, VIC 3207, Australia

314–321, 3rd Floor, Plot 3, Splendor Forum, Jasola District Centre, New Delhi – 110025, India

103 Penang Road, #05–06/07, Visioncrest Commercial, Singapore 238467

Cambridge University Press is part of Cambridge University Press & Assessment,
a department of the University of Cambridge.

We share the University's mission to contribute to society through the pursuit of
education, learning and research at the highest international levels of excellence.

www.cambridge.org
Information on this title: www.cambridge.org/9780521711975

First published 2009

A catalogue record for this publication is available from the British Library

Library of Congress Cataloging-in-Publication data
Pyke, Kevin.
Plastid biology / Kevin Pyke.
p. ; cm.
Includes bibliographical references and index.
ISBN 978-0-521-88501-0 (hardback) – ISBN 978-0-521-71197-5 (pbk.)
1. Plastids.   I. Title.
[DNLM: 1. Plastids.   2. Molecular Biology–methods.
3. Plant Physiology.   QU 350 P995p 2009]
QK725.P95 2009
571.6′59–dc22

2008045753

ISBN   978-0-521-88501-0   Hardback
ISBN   978-0-521-71197-5   Paperback

Contents

Preface   *page* vii
Acknowledgements   viii

1. **What are plastids and where did they come from?**   1
2. **Different types of plastids and their structure**   9
3. **The plastid genome – structure, transcription and translation**   31
4. **Photosynthesis**   61
5. **Plastid import**   81
6. **The development of the chloroplast**   106
7. **Plastid metabolism**   130
8. **Plastids and cellular function**   153
9. **Plastid transformation and biotechnology**   178

Further reading and resources   195
Index   197

Preface

Plants are fundamental in enabling the planet Earth to function as a relatively stable system. They exert control over the biosphere by their interaction with their environment, a fundamental aspect of which is fixing carbon dioxide from the atmosphere and generating oxygen, in the process of photosynthesis. This critical process is carried out by specialised chloroplast organelles within green plant tissues, primarily leaves. However, chloroplasts are only one member of a family of organelles called plastids, which reside in all plant cells, and which take on different forms in relation to their cellular function, biochemistry and storage capacity. For many years, photosynthesis research overshadowed other aspects of plastid biology, but in the last two decades, much new knowledge about how plastids function and how they relate to their evolutionary past has become available from research. This book provides an overview of a wide range of aspects of modern plastid biology, including a consideration of different plastid types and how they relate to cell function, plastid genomes and how proteins are imported into plastids, photosynthesis and core aspects of plastid biochemistry, plastid signalling and functionality within a cellular context and plastid genetic manipulation. The modern era of molecular biology and molecular genetics has enabled much to be learnt about how plastids function and a picture is revealed of a highly complex organelle at the very core of plant cellular function. This information should be useful for final-year undergraduate students or Masters students interested in plant sciences and cell biology.

Acknowledgements

I would like to thank the many people with whom I have collaborated over the years and learnt about plastids in many different guises. Firstly, Rachel Leech, in whose laboratory I first became properly acquainted with plastids and, in more recent times, Anil Day, Rupert Fray, John Gray, Julian Hibberd, Enrique López-Juez, Simon Møller and Kathy Osteryoung, to name a few. Thanks also to those who have provided material for figures in this book and a special thank you to Mark Waters for his skills in drawing the figures. I would also like to thank Cambridge University Press for providing the opportunity to bring together a volume of this nature.

1

What are plastids and where did they come from?

Plastids are a group of organelles present in the cells of all higher and lower plants, including algae, which function in a variety of different ways to enable plants to grow and function. Although different types of plastids which are found in different types of cells have modified roles, according to the type of cell in which they reside, the foremost function of plastids is carrying out the process of photosynthesis. Photosynthesis is a fundamental feature of plants and is facilitated by the presence of green, pigmented chloroplasts within plant cells. Indeed, photosynthesis is a defining feature of plants and enables them to fix carbon from the gaseous carbon dioxide in the air and synthesise a variety of complex organic molecules which allows them to increase in stature and mass. Photosynthesis is carried out by chloroplasts, which by virtue of containing the green pigment chlorophyll, defines the phenotype of green plants. Photosynthetic Eukaryotes have increased in their complexity dramatically since the first land plants, termed Embryophytes, evolved from freshwater multicellular green algae, around 450 million years ago. The current-day group of algae that are most closely related to these ancient algae are the Chlorophytes (Fig. 1.1). From these have evolved the lower plants, which includes the liverworts, mosses, hornworts and ferns (Fig. 1.1). Subsequently, the Gymnosperms and then the flowering plants, the Angiosperms, evolved and the Angiosperms, in particular, have been highly successful in conquering the planet such that much of the Earth is covered in green swathes of vegetation containing countless numbers of photosynthetic chloroplasts within their cells. Although extensive research on plastid biology has been done on lower plants, the majority has been carried out using higher plants and particularly members of the Angiosperms, which include all of the world's crop plants. Thus, in this book the focus will be on an understanding of plastid biology in

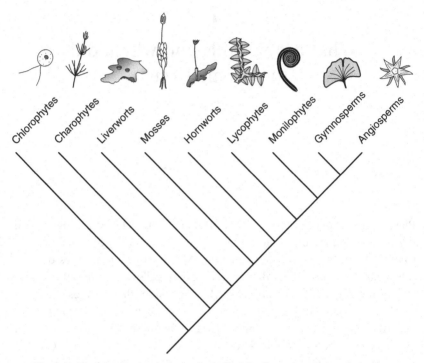

Fig. 1.1. Modern-day land plants, including the flowering plants or Angio-sperms, evolved from an ancestral green alga which closely resembled today's Chlorophytes. This gave rise to the multicellular Charophyte algae, the most closely related relatives of land plants. Following invasion of the land, several distinct groups of land plants subsequently evolved: the liver-worts, mosses, hornworts, Lycophytes (club mosses) and the Monilophytes (ferns). The most derived land plants bear seeds, and are represented by the Gymnosperms and the Angiosperms (the flowering plants).

Angiosperms rather than in lower plants. Although chloroplasts are the major form of plastids in Angiosperm plant cells, other plastid types have evolved to take up modified roles in different types of cells and tissues, especially in relation to storage of different types of molecules. Different types of plastids and their roles are considered in detail in Chapter 2.

A major question therefore regarding the evolution of higher plants is from where photosynthetic plastids arose, since they are fundamental to the successful evolution of higher plants. The source of plastids in modern-day plants has been a much-debated subject for a long time. Andreas Schimper in 1883 first suggested that chloroplasts were a result of a symbiosis between a photosynthetic organism and a non-photosynthetic host, and this outrageous suggestion was further supported by the Russian botanist Constantin Mereschkowsky in 1905. Much debate continued through the twentieth century but was refocused as a hypothesis based on microbiological observations by Lynne Margulis in 1967 in her paper 'on the origin of mitosing cells' (*Journal of Theoretical Biology*, **14**, 255–274). At the time, this hypothesis was heavily criticised and Margulis showed great persistence in sticking with what was an unorthodox hypothesis, and which in the modern era of plant biology, with the availability of genome sequences from both the plant nucleus and from the genome present within the plastids (see Chapter 3), has become the generally accepted hypothesis of how plastids evolved. As such, it has become clear that plastids have a prokaryotic history, in that they most likely evolved originally from a free-living photosynthetic prokaryotic organism. Although the precise organism from which plastids evolved may no longer be found on the planet, through the course of its own evolution, the closest living modern-day group are the Cyanobacteria, the photosynthetic bacteria (Fig. 1.2). Cyanobacteria are free-living single-celled prokaryotic organisms, sometimes aggregated into clusters or filaments and capable of carrying out photosynthetic carbon fixation. Although they use the photosynthetic light absorption pigment chlorophyll, they also use other pigments, such as phycoerythrin and phycocyanin, and thus Cyanobacteria are so named due to their colour. Cyanobacteria are abundant across the planet and play a major role in global photosynthesis as well as in the fixation of atmospheric nitrogen. They are found largely in aqueous habitats, or at least in habitats which are damp and have rather confusingly been referred to extensively in the past as blue–green algae, even though they are not algae but bacteria. So how did Cyanobacteria give rise to chloroplasts found inside cells of modern-day higher plants?

The modern theory of endosymbiosis, as first put forward by Margulis, suggested that a free-living photosynthetic prokaryote was engulfed by phagotrophy and taken up into a proto-eukaryote cell, within which it evolved a symbiotic relationship with the host cell (Fig. 1.3). Margulis used various strands of evidence gathered by microscopical examination of chloroplasts, but a central element was the discovery that plastids contained their own DNA, which form their own plastid genomes. In addition, a significant number of features characteristic of modern-day plastids clearly arose from their prokaryotic history and many of these features are discussed in detail in other chapters in this book.

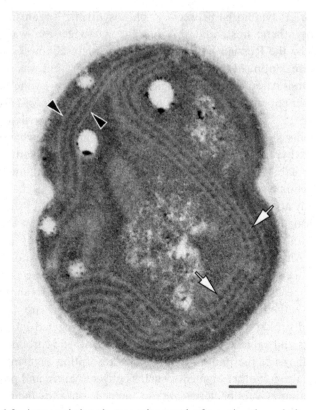

Fig. 1.2. A transmission electron micrograph of a section through the cell of a Cyanobacterium, *Synechocystis* sp. PCC6803. The thylakoid membranes within the cell (black arrows) and the light harvesting complexes, termed phycobilosomes (white arrows), are visible. Scale bar = 1 μm. (Image courtesy of Conrad Mullineaux, Queen Mary, University of London.)

Foremost amongst these features are:

1. Plastids contain their own DNA, but much of that DNA has moved into the cell's nuclear genome through the course of evolution and is now missing from the plastid genome. Characteristics of plastid DNA structure and its molecular biology show many similarities with prokaryotic genomes from bacteria.
2. Proteins encoded by genes, which were originally plastid-encoded, and which are now present in the nucleus, are imported into the plastid after translation on ribosomes in the cytoplasm of the cell.
3. Plastids in higher and lower plants are surrounded by a double membrane, the inner membrane of which has features similar to

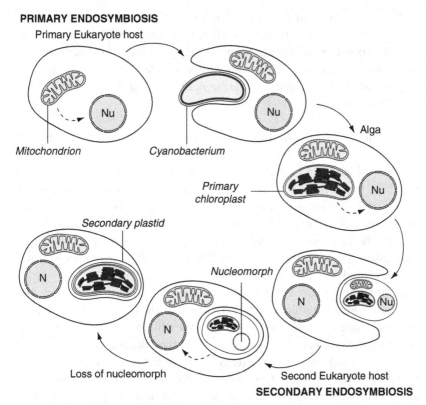

Fig. 1.3. The proposed origin of plastids in higher and lower plants occurred by a process of phagotrophy and the establishment of an endosymbiotic relationship. A photosynthetic Cyanobacterium was taken up by the primary eukaryotic host and incorporated into a stable endosymbiotic relationship, giving rise to a eukaryotic algal-like organism, containing mitochondria, a primary chloroplast and a nucleus (Nu). Subsequent secondary endosymbiotic events, in which an entire alga was taken up by another eukaryotic host cell also occurred. In this case, the algal nucleus forms the nucleomorph (Nu), which becomes reduced in size. N indicates the nucleus of the second eukaryote host. Broken arrows indicate gene transfer from organelles to the host nucleus. (Redrawn from Larkum WD, Lockhart PJ, Howe CJ., Shopping for plastids, *Trends in Plant Science* 12, 189–195. © Elsevier 2007.)

that of bacteria membranes and which is different to other cellular membranes.

4. The internal structure of the plastid, especially the thylakoid membrane and its protein complexes found in chloroplasts are similar to that found in Cyanobacteria.

5. Plastids contain their own ribosomes, which have many features linking them with prokaryotic ribosomes and are different to those eukaryotic-type ribosomes found in the cytoplasm of the cell.
6. Plastids are capable of division within the plant cell and divide by a process of binary fission, similar in many ways to the division process of bacteria.

The theory of endosymbiosis has been supported extensively by the avalanche of new information about genes and proteins functional in plastids, which have been discovered in the last 20 years, especially as a result of the complete sequencing of nuclear and plastid genomes in plants, and it is now generally accepted that this was how plastids found in modern-day higher and lower plants originated. In addition, a previous endosymbiotic event involving non-photosynthetic bacteria is believed to have given rise to mitochondria, since they also show many prokaryotic features, similar to plastids.

In order for a successful endosymbiotic relationship to succeed, the symbiont has to be able to divide itself within the cytoplasm of the host cell and synchronise such divisions with the cell division of the host cell, such that the symbiont is maintained within the cells. Subsequent integration appears to have required extensive movement of genetic information to the nucleus, evolution of mechanisms to re-import those gene products along with the hijacking of nuclear genes for control of plastid function, and loss of the peptidoglycan cell wall from the symbiont.

A key question in the evolution of plastids is how many times did this original endosymbiotic relationship occur to give rise to the wide variety of plastid-containing organisms found on the planet today. Although this question has been much debated, it seems that the most likely scenario is that such a primary endosymbiosis occurred once and gave rise to three distinct groups of Eukaryotes, the Green algae, the Red algae and the Glaucophytes (Fig. 1.4). The precise order in which these groups have arisen through evolution is still open to significant debate, as is the discussion as to whether this original endosymbiotic event was a unique event or whether it occurred more than once. From this original event, the three groups of Red algae, Green algae and Glaucophytes have evolved into a large collection of organisms, consisting of around 15000 different species. Most of these are single-celled aquatic organisms but some have evolved significant multicellular structures and differentiated tissues, most notably in the seaweeds. From the Green algae eventually arose the evolutionary lineage, which gave rise to the first land plants (Fig. 1.1).

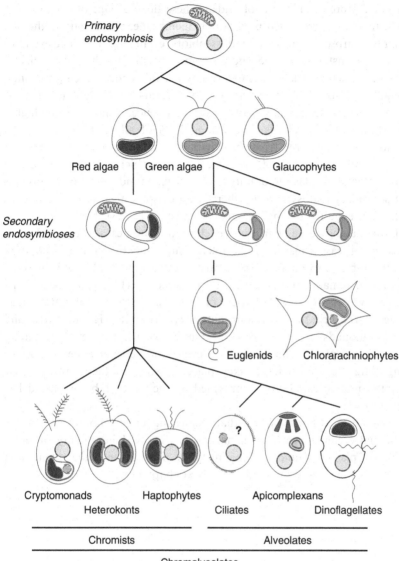

Primary endosymbiosis

Red algae   Green algae   Glaucophytes

Secondary endosymbioses

Euglenids   Chlorarachniophytes

Cryptomonads   Haptophytes   Apicomplexans

Heterokonts   Ciliates   Dinoflagellates

Chromists   Alveolates

Chromalveolates

Fig. 1.4. A diagram of plastid evolution in Algae. Primary endosymbiosis took place once to yield Glaucophytes, Red algae and Green algae, which eventually gave rise to land plants and Angiosperms. Secondary endosymbiosis involving Green algae occurred twice, giving rise to Euglenids and Chlorarachniophytes. A single secondary endosymbiosis involving a Red alga gave rise to the many groups called Chromalveolates. In many of these groups, the ability to carry out photosynthesis has been lost. (Redrawn from *Protist* 155, Keeling P, A brief history of plastids and their hosts 3–7. © Elsevier 2004.)

A key feature of all of the plastids in these three groups of organisms is that they are surrounded by a double membrane, which reflects the fact that they arose by a primary endosymbiotic event (Fig. 1.3). The extent of genetic transfer from the endosymbiont plastid to the cell's nucleus has been massive, such that modern-day plastid genomes are significantly smaller in size than those found in current-day Cyanobacteria (see Chapter 3). Indeed, analysis of the fully sequenced genome of the higher plant *Arabidopsis* reveals that around 18% of all genes in its nuclear genome were acquired from the original prokaryotic endosymbiont.

A significant aspect of plastid evolution, which has resulted in the diverse array of plastid-containing organisms found today, is the process of secondary endosymbiosis whereby an existing algal cell containing a plastid is engulfed by another eukaryotic cell (Fig. 1.3). This results in the plastid having more than two envelope membranes as well as the cell having, at least transiently, two nuclei. Subsequently, in most secondary endosymbiotic events, the secondary nucleus is lost, although in some groups it remains and is termed a nucleomorph (Fig. 1.3). Such events have given rise to a wide variety of organisms evolved from Red algae (Fig. 1.4), including Dinoflagellates, Heterokonts, Haptophytes and Cryptomonads. In such cases, the number of membranes surrounding the original plastid can be three or even four, reflecting the number of engulfing events that have taken place. In some of these groups, their photosynthetic capability as provided by the endosymbiont plastid has been lost.

What is most remarkable is that the original endosymbiont, which was essentially chloroplast-like in nature, has evolved into many different types of plastids found in different types of tissues in higher plants and these are considered in detail in the next chapter.

2

Different types of plastids and their structure

Since the first endosymbiotic event that enabled a free-living photosynthetic organism to take up residence in a eukaryotic cell, the organisms that we know of as plants have evolved in a dramatic and highly varied way. In particular, plants have evolved from their single-celled ancestors into complex multicellular structures. A key characteristic of these multicellular plants is that they contain cells of different types, which are distinguishable from each other in the functions that they perform within the whole complex organism. With increasing complexity of form through the evolution of Bryophytes, Monilophytes and into the higher plants, Gymnosperms and Angiosperms, large numbers of different types of cells have been developed such that, in the more complex members of the flowering plants, the Angiosperms, there are over 50 different types of cells. As a result of this diversification of cell types in plants, the original endosymbiotic plastid found itself being manipulated by the host cell to take on a variety of different roles in these different types of cells. Plastids evolved from their original photosynthetic function after the original endosymbiosis to take up a key role in the cell as a whole, particularly in relation to biochemical interactions in the cell's metabolism. As a result, in modern-day higher plants, there are a variety of different types of plastids, which fulfil different roles in different types of plant cells. The situation, however, is not clear-cut. There is significant interaction between different plastid types in different types of cells in that the plastids can interconvert between different types according to molecular and environmental signals. In addition, the plastid populations in many types of cell are not entirely homogeneous and may be composed of a mixture of different types of plastids. Here we will consider the basic types of plastids found in higher plants.

Proplastids

All of the cells in a plant are derived from cells produced in highly organised areas of tissue called meristems. Meristems are found at the tips of shoots and at the tips of roots and act as a source of undifferentiated cells, which are recruited from the meristem by cell division and cell expansion to form complex organs such as leaves, roots and flowers. During the formation of such organs, these undifferentiated cells differentiate into distinct cell types and contribute functionality to the organ in question. All of the plastids in all the cells in a plant are derived from those plastids found in the meristem cells within that plant and these progenitor plastids are called proplastids. Proplastids are essentially undifferentiated plastids, and are found extensively in root and shoot meristems, as well as in other tissues such as cells in early embryo development in seeds and other tissues which contain young dividing cells. Proplastids are also present in cultured plant cells and callus tissues on cultured plant explants, which are non-green. Most of the knowledge about proplastids has come from observing sections of meristem tissues with the electron microscope (Fig. 2.1), since proplastids are small, about 1–2 μm in length and contain little, if any, pigment. Thus they are difficult to study with conventional light microscopy. They can be viewed more effectively, however, by introducing into them a fluorescent molecule, and then observing them using fluorescence or confocal microscopy (Fig. 2.2). As with all plastids, proplastids are surrounded by a boundary double membrane, called the plastid envelope and, internally, they contain small pieces of a membrane system called the thylakoid membrane (Fig. 2.1), which becomes considerably more extensive in other differentiated types of plastid. Plastids resident in the shoot apical meristem contain a more organised array of thylakoid membrane than those in the root apical meristem. In addition to thylakoid membrane, proplastids contain their own ribosomes and, in seeds, proplastids often contain grains of starch, which are laid down during seed development and form a source of nutrition during seed germination. In specific tissues such as the plumule of wheat seedlings and in the stolon tissues of potato, many proplastids contain starch grains whilst others contain none. This difference relates to the presence or absence of a key starch biosynthetic enzyme, starch synthase, in different subgroups of proplastids within a cell.

The extent of variability of proplastids between meristem cells or within the same cell is difficult to examine because of the small size of the meristem cells and the dense meristem tissue. However, there is some

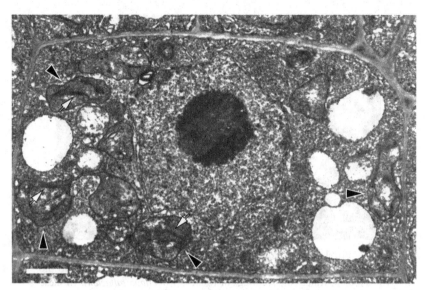

Fig. 2.1. Transmission electron micrograph of a sectioned cell from the shoot apical meristem of a seedling of *Arabidopsis thaliana*. Several proplastids are present (black arrows), defined by an outer membrane and containing small sections of internal thylakoid membrane (white arrow). The proplastids surround a large nucleus in the centre of the cell, containing a darkly stained nucleolus. Several white vacuoles are also present. Bar = 1 μm.

variation in proplastid number per cell between different areas of the shoot apical meristem, varying between 10–20 proplastids per cell. Since proplastids are found in tissues in which cell division takes place, a fundamentally importantly aspect of proplastid biology is their own division and how they segregate into the two daughter cells in the latter stages of cell division, at cytokinesis. Failure to divide or segregate in the correct manner would quickly lead to meristem cells with no proplastids, which, with subsequent divisions, would lead to large numbers of aplastidic cells, containing no plastids. Since plastids are crucial for many aspects of basic cellular metabolism, this condition would likely be lethal and thus meristems and organs derived from such cells would quickly die. Proplastids divide by constriction in the centre of the plastid, leading to a pinching off of two new daughter proplastids in a process termed binary fission. This process occurs in other types of differentiated plastids and is discussed in detail in Chapter 8. As a result of proplastid division prior to cytokinesis and the segregation of proplastids into the new daughter cells, proplastid numbers are maintained consistently in meristem cells.

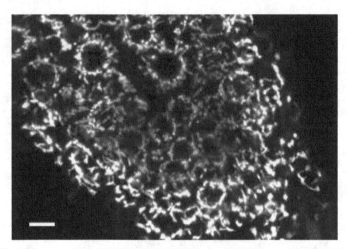

Fig. 2.2. Proplastids visualised in cells in the region of the root meristem by using green fluorescent protein (GFP), which is genetically targeted to the proplastids in transgenic seedlings of tobacco (*Nicotiana tabacum*). In these plants, all the proplastids in these cells fluoresce green and are shown here, in grey scale, as pale organelles within individual cells. Bar = 10 μm. (From Kohler RH, Hanson MR (2000). Plastid tubules of higher plants are tissue-specific and developmentally regulated. *Journal of Cell Science* 113, 81–89. Reproduced with permission from the Company of Biologists.)

The most likely strategy by which proplastids are segregated relatively equally into the two daughter cells is as a result of their even distribution throughout the cell's cytoplasm, ensuring that a central plane of cell division will always result in proplastids in both daughter cells. It may be that a more directed positioning of proplastids close to the nucleus and surrounding it (Fig. 2.1), as directed by the cell's cytoskeleton is also possible, thereby also ensuring equality of segregation into the two newly formed cells.

Another important aspect of proplastid biology is their role in plastid inheritance between generations. During the formation of pollen and egg cells in flowers, proplastids are present. However, in many species of Angiosperms, proplastids are excluded from or degraded during pollen development, such that at fertilisation and the formation of the zygote, only proplastids from the maternal line are present. Thus, in the majority of Angiosperms, plastids are maternally inherited. Consequently, all of the plastids in the subsequent generation are derived from the maternal line and not the paternal line. In about one-third of Angiosperm species, some degree of biparental inheritance of proplastids occurs, such that some plastids are present in pollen and thus the cells of the zygote can contain a mixture of maternal and paternal plastids. A variety of factors can dictate the actual proportion of maternally and paternally derived plastid types in the resulting plants, and often early sorting out of the two types of plastid at cell division results in the mature plant containing predominantly one type of inherited plastid. In contrast, in the Gymnosperms, the conifers show predominantly paternal plastid inheritance, whereas other Gymnosperm groups are similar to the Angiosperms in having maternal plastid inheritance. Exactly why plants have evolved these mechanisms, which ensure that only one plastid type is predominantly inherited, is unclear.

Chloroplasts

In many of the cells of the above-ground tissues of plants, proplastids differentiate into green, pigmented plastids called chloroplasts. These are the primary type of plastid present in all green parts of plants by virtue of their green pigment, chlorophyll. Typically, these tissues are leaves but also include stems, petioles, pods, immature petals, sepals, tendrils and a variety of other green plant structures. Chloroplasts carry out the process of photosynthesis, which is described in more detail in Chapter 4, and many of the other detailed aspects of chloroplast biology are considered in the other chapters in this book. Compared to other plastid types, the chloroplast is by far the best understood in terms of its biology, its biochemistry and molecular biology and how it interacts with the rest of the cell.

Palisade and spongy mesophyll cells in leaves are the major sites for chloroplast development and they accumulate to form populations of significant size in these cells by dividing (see Chapter 8). Consequently, in leaves of most plants, the mesophyll cells contain between 50 and 200 chloroplasts, depending on the size of the cell and the size of the individual chloroplasts. Since these mesophyll cells are highly vacuolated, the chloroplasts form a monolayer within the cytoplasm of the cell, pushed against the cell wall by turgor pressure of the vacuole. When mesophyll cells are fixed and separated and viewed microscopically, the layer of chloroplasts on the upper surfaces of the cell can be seen. Using Nomarski optics, the chloroplasts on the lower surfaces of the same cell can be seen also (Fig. 2.3). Such an image clearly shows how packed with chloroplasts mesophyll cells really are. In terms of coverage of the

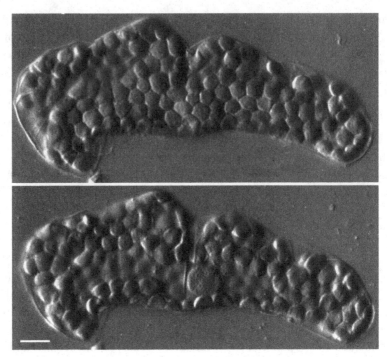

Fig. 2.3. A mesophyll cell from a leaf of wheat (*Triticum aestivum*), which has been fixed and separated from other cells and observed with Nomarski optics, which allows the upper and lower surfaces of the cell to be seen separately. The chloroplasts fill most of the cell's cytoplasm and are tightly packed together and those individual chloroplasts on the upper (top panel) and lower (bottom panel) cell surfaces are clearly seen. A circular granular nucleus can be seen in the middle of the cell in the bottom panel. Bar = 10 μm.

available surface area of the mesophyll cell, chloroplasts occupy up to 70% of the total surface area. In leaves grown in moderate light levels, the chloroplasts do not tend to be located in areas of cytoplasm internal to cell walls which are joined to a neighbouring mesophyll cell, which accounts for about 30% of total surface area. So in effect, chloroplasts in mesophyll cells generate a population, which covers entirely the internal surface area of the mesophyll, which is exposed to airspaces within the leaf's three-dimensional internal architecture. This obviously optimises the efficiency of gaseous exchange between airspaces and the chloroplast, since it minimises the pathway for gaseous diffusion. Images of sectioned mesophyll cells classically show sectioned

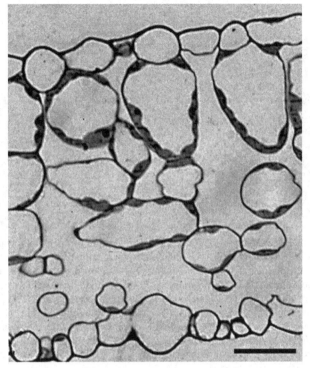

Fig. 2.4. Transverse section through a leaf of *Arabidopsis thaliana*, showing how the chloroplasts in the mesophyll cells form a monolayer, pressed against the cell wall by the large vacuole, which takes up most of the cell's volume. Chloroplasts are present in the epidermal layers of cells, at the upper and lower surface of the leaf but are less well developed. Bar = 10 μm.

chloroplasts as small bumps on the inside of the cell wall, distorting the tonoplast membrane and pressed tightly against the cell wall by the vacuole (Fig. 2.4).

Chloroplasts vary in size somewhat between different species but also between different cell types within an organ. For instance, in a leaf, the chloroplasts in the epidermal cells covering the leaf surface are significantly smaller and poorly developed compared with mesophyll chloroplasts, but do contain low levels of chlorophyll and should be considered as chloroplasts. In many texts, it is stated that epidermal cells lack chloroplasts, which is untrue. Chloroplasts in the guard cells, which form stomata in the epidermis are well developed, although smaller than mesophyll chloroplasts. Likewise, those chloroplasts present in the bundle sheath cells surrounding the vascular bundles in the leaf are also well

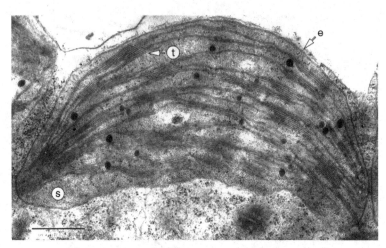

Fig. 2.5. Electron micrograph of a transverse section through a chloroplast from a leaf mesophyll cell. The outer envelope membrane can be seen as a double membrane (arrowed e) and internally the chloroplast has an extensive thylakoid membrane system (arrowed t), which resides in the chloroplast's stroma (labelled s). The numerous round dark bodies are arrays of lipid molecules and protein and are called plastoglobuli. Bar = 1 μm. (From Pyke K, Waters M (2005). Plastid development and differentiation. In *Plastids* (ed. Møller SG). *Annual Plant Reviews* 13. Reproduced with permission from Wiley-Blackwell Publishing.)

developed but small. This difference is exacerbated in leaves of some grass species, which carry out the specialised type of photosynthesis termed C_4. In these grasses, such as maize, there is a clear difference in development and size between the chloroplasts in the bundle sheath cells and the chloroplasts in the surrounding mesophyll cells (see Chapter 8).

The structure of the chloroplast is optimised for increased efficiency of photosynthesis by having an extensive internal membrane system, termed the thylakoid membrane (Fig. 2.5). In the thylakoid membrane reside many protein complexes containing bound chlorophyll molecules, which are part of the photosynthetic machinery involved in light capture and energy transduction. Consequently, a large area of thylakoid membrane within the chloroplast enhances chlorophyll content and light absorption. Classically, chloroplasts are supposed to resemble an oblate spheroid in their three-dimensional shape, similar in shape to a flattened rugby ball or American football, although this structure is highly variable, and chloroplast morphology is much more dynamic than static images might suggest (see Chapter 8). Chloroplasts viewed by electron microscopy or as

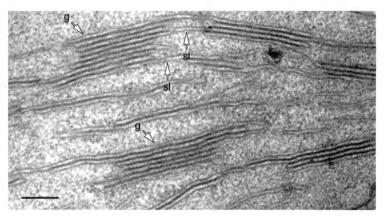

Fig. 2.6. An electron micrograph of the chloroplast thylakoid membrane. This image is a transverse section through the complex three-dimensional network of thylakoid membrane. The granal stacks of appressed thylakoid membrane (arrowed g) are linked by stretches of stromal lamellae (arrowed sl). The thylakoid membrane contains an internal space between the two membrane surfaces called the thylakoid lumen, seen as light areas between darkly stained membrane. The granular area surrounding the thylakoid membranes is the stroma. Bar = 250 nm. (Image courtesy of Mark Waters, Plant Sciences, University of Oxford, UK.)

isolated organelles clearly take up this shape (Fig. 2.5). A consideration that chloroplasts have distinct poles at either end, which relates to plastid function, was not made until proteins involved in the plastid division mechanism were shown to locate precisely at either pole during division (see Chapter 8). Thus, the chloroplast should be considered a polarised organelle with two poles and an equator at the middle. The average plan shape of chloroplasts is between 5 and 10 μm in length and 2 and 5 μm in width, although this can vary extensively between chloroplasts in different types of cells and between species. Thus, their average plan area is about 50 μm². Within the chloroplast, the extensive thylakoid membrane system is estimated to cover around 500 μm². Extrapolating this to an average leaf, in which the mesophyll and palisade cells contain around 100 chloroplasts, the total area of thylakoid membrane is estimated to be between 300 and 800 m² per m² of leaf surface, representing a highly significant level of packing efficiency. The thylakoid membrane forms a complex architecture of stacked arrangements called grana (Fig. 2.6), which form as a result of electrostatic interactions between protein complexes on the thylakoid surface. Thus, the architecture of the thylakoid should be considered dynamic and not fossilised in a permanent arrangement as electron micrograph images would tend to show (see Chapter 6). Areas of unstacked thylakoid membrane, which link granal stacks together, are called stromal lamellae (Fig. 2.6). The way in which the thylakoid membrane is constructed is discussed in detail in Chapter 6. The thylakoid membrane is composed as a double membrane and thus forms a distinct compartment within it called the thylakoid lumen (Fig. 2.6). This lumen plays an important role in various aspects of photosynthesis, especially in facilitating a proton gradient across the thylakoid membrane that enables production of ATP (see Chapter 4).

The remainder of the chloroplast's internal volume, which surrounds the thylakoid membrane, is called the stroma, in which resides a variety of enzymes and protein complexes, complexes of DNA molecules associated with inner envelope membrane, plastid ribosomes, accumulations of starch in the form of granules and lipid bodies called plastoglobuli (Fig. 2.5). The stroma has been referred to as the liquid phase of the chloroplast but the concentration of molecules and complex structures within the stroma is such that one should consider it more like a gel matrix than an aqueous fluid. Indeed, in some circumstances, the concentration of the photosynthetic carbon fixation enzyme ribulose 1,5 bisphosphate carboxylase (RUBISCO) in the stroma is such that crystalline arrays of the enzyme can occur. Plastids contain their own ribosomes, which reflects their prokaryotic prehistory, and in chloroplasts large populations of ribosomes can be observed in electron micrographs of the stroma. Estimates suggest that chloroplasts contain around 10^5 ribosomes, which are involved in protein translation of mRNA molecules, derived from transcription of genes on the plastid DNA (see Chapter 3). In many plants, chloroplasts accumulate grains of starch towards the end of the day, which represent the end product of photosynthetic carbon fixation through that day. During the following night, the starch is broken down and exported into the cytosol. In contrast, some plants are not starch accumulators and export their fixed carbon through the light period. Starch grains can be large and numerous and can distort the morphology of the chloroplast significantly. Chloroplasts vary extensively in size and complexity of their thylakoid membrane between different cell types and between different species. They also have the ability to differentiate into different types of plastid, mainly by virtue of changing the type of, and extent of, their primary storage material. Some of these distinct types of other plastids will now be considered.

Amyloplasts

The ability of the chloroplast to accumulate transitory grains of starch during the day has been extended in the amyloplast, which is a distinct type of plastid in which starch granules are laid down in a long-term storage mode. Amyloplasts; *amylo* meaning starch, are the site of long-term starch storage in plants and are the primary plastid type found in storage tissues such as the endosperm in seeds and tubers and storage organs associated with roots (Fig. 2.7). All of the starch extracted from the world's crops such as grains and tubers was originally laid down in amyloplasts within the cells of storage tissues. This starch is synthesised either directly from photosynthate, as occurs in leaves or indirectly from photosynthate transported into heterotrophic storage tissues, which is the process by which most amyloplasts form. Consequently, amyloplast biology and the way in which the starch is accumulated in amyloplasts are of major importance from a food and nutrition point of view, since 75% of energy in the average human diet is derived from starch.

Most amyloplasts arise directly from proplastids during development of the storage tissues. In the endosperm of wheat seeds, proplastids are present within the coenocytic endosperm early in development, but after the onset of cellularisation, proplastids start to accumulate starch granules

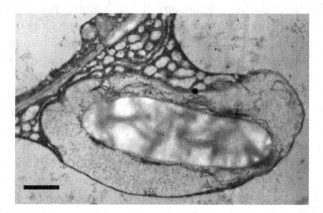

Fig. 2.7. An electron micrograph of a sectioned amyloplast from the cotyledon of a developing seed of pea (*Pisum sativum*). The section shows a single large grain of starch within the amyloplast with some detail of surface structures which surround it. Bar = 10 μm. (From Pyke K, Waters M (2005). Plastid development and differentiation. In *Plastids* (ed. Møller SG). *Annual Plant Reviews* 13. Reproduced with permission from Wiley-Blackwell Publishing and courtesy of Alison Smith, John Innes Centre, Norwich, UK.)

and amyloplasts are formed. Surprisingly, considering their importance, an understanding of the molecular controls that govern amyloplast differentiation in storage tissues is limited and most of the research on this differentiating system has come from inducing amyloplast formation in cultured tobacco cells. Amyloplasts will differentiate from proplastids in these cells when the hormone auxin is replaced with the cytokinin hormone, benzyladenine, which reduces rates of cell division and induces starch accumulation and amyloplast formation. Whether this is the basis for the way in which amyloplasts form in vivo plant tissues is unclear. Although amyloplasts have long been considered solely a compartment for starch storage, more recent analysis of the proteins and biochemical capabilities of isolated amyloplasts have shown that they contain a broad range of enzymes and molecular machinery similar in many ways to that of the chloroplast, although lacking significant thylakoid membrane containing chlorophyll, and hence they lack photosynthetic metabolism.

Starch itself is an insoluble semi-crystalline polymer of glucose synthesised in the amyloplast by the polymerisation of ADP glucose (see Chapter 7). This produces two types of molecule: amylopectin, which is highly branched, and amylose, which is a linear polymer. These two starch molecules generally accumulate in a proportion of 70:30%, respectively in a starch granule. Within a starch granule, there is a highly organised structure, which relates to the organisation of the amylopectin chains. Scanning electron microscopy of starch grains from amyloplasts shows them to be approximately spherical and reveals a series of concentric rings, which reflect differences between semi-crystalline and amorphous zones in the way that the amylopectin chains are arranged (Fig. 2.8). In addition, there are differences in the overall size of starch granules, which form different size classes. In cereal endosperm amyloplasts, type A granules are up to 45 μm in diameter and type B grains are up to 10 μm in diameter (Fig. 2.9) but in general, starch grain sizes in amyloplasts can vary a hundred-fold between 1 and 100 μm in diameter. These large and small starch granules are normally synthesised in different amyloplasts, although variation in starch granule size within a single amyloplast does occur. There are distinct differences between starch granules in amyloplasts and those starch grains laid down in a transitory form during the day in chloroplasts in leaf cells. In transitory starch grains, the amylose content is lower and the starch forms as a flattened plate-like structure, lacking the growth rings of starch granules in amyloplasts. Transitory starch can take on properties of long-term starch in that, in older leaves of some plants, such as tobacco and cotton, starch accumulated during the

Fig. 2.8. Scanning electron micrograph of a starch grain from an amyloplast in a tuber of potato (*Solanum tuberosum*) showing how the starch grain is built of concentric rings, reflecting differences in the organisation of amylopectin chains. The internal crack is an artefact of sample preparation. Bar = 5 μm. (Image courtesy of Alison Smith, John Innes Centre, Norwich, UK.)

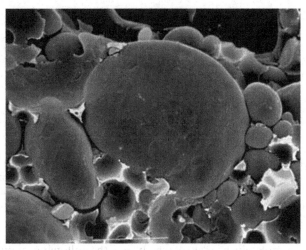

Fig. 2.9. Scanning electron micrograph of starch grains contained within amyloplasts in the endosperm of wheat seeds, showing significant variation in size between individual grains. The large grain in the centre shows indentations on its surface where smaller grains were pressed before the sample was prepared. The envelope membrane of the amyloplast which surrounds these starch grains is not visible. (Image courtesy of Joseph R. Thomasson, Professor of Botany, Department of Biological Sciences, Elam Bartholomew Herbarium, Fort Hays State University, USA.)

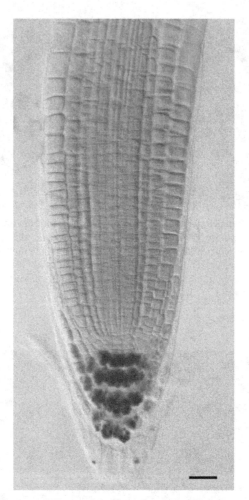

Fig. 2.10. The root tip from a seedling of *Arabidopsis thaliana* stained with iodine and revealing starch as blue/black granules in the statoliths of the columella cells at the root tip. Bar = 40 μm. (Image courtesy of Ranjan Swarup and Malcolm Bennett, Division of Plant and Crop Sciences, University of Nottingham, UK.)

day is not completely broken down during the following night and so accumulates in the leaf chloroplasts on a more long-term basis.

During the evolution of gravity-sensing mechanisms in plants, the ability of relatively dense amyloplasts to sink in the cytoplasm under the influence of gravity has been exploited as the basis for gravity sensing. In roots, gravity sensing is carried out by the columella cells, which reside just behind the root cap (Fig. 2.10). They contain specialised amyloplasts, which in these columella cells are called statoliths. Statoliths sink to the bottom of the columella cell under the influence of gravity and initiate a signal transduction pathway involving redistribution of the plant hormone auxin, resulting in differential cell expansion on either side of the root and the downward growth of the root in a positively gravitropic manner, that is toward the centre of the earth. In stems of shoots, cells surrounding the vascular tissues also contain statoliths and invoke the upwards directional growth of shoots, which are negatively gravitropic.

Chromoplasts

Many Angiosperms make use of animals to assist with pollination and seed dispersal. With the evolution of fleshy fruits to facilitate seed dispersal, a requirement for some type of attractant strategy arose. Thus, brightly pigmented structures arose on Angiosperms and the plastid was hijacked as the site of storage for many of these novel pigment molecules. These specialised types of plastids are called chromoplasts and are the site of pigment accumulation in cells of flower petals, fleshy fruits and in a variety of other pigmented structures found in the diverse array of flowering plants. Chromoplasts are not the sole site of pigment storage since many coloured molecules, such as anthocyanins, can accumulate in the cell's vacuole, and indeed in many petals, the final colouration is a combination of chromoplast pigments and vacuolar pigments. In most cases, chromoplasts develop from chloroplasts and this process is classically observed during the ripening process of some fleshy fruits, most notably that of the tomato (*Solanum lycopersicon*), where unripe green fruit containing chloroplasts turn into ripe red fruit containing chromoplasts. The main class of pigment molecules, which accumulate in chromoplasts, are carotenoids, which are synthesised from the C_{40} molecule phytoene and of which there are many different types with differing colours (see Chapter 7). For instance, carotenes are orange, lycopene is red, and zeaxanthin and violoxanthin are yellow. Several of these carotenoid molecules are also found in complexes on the thylakoid membrane of the chloroplast, where they act as accessory pigments in light capture and energy dissipation by the chlorophyll antenna complexes (see Chapter 4). Pigmented tissues normally contain a mixture of carotenoids, which define their overall colouration. For instance, ripe tomatoes are red and their chromoplasts contain lycopene and β-carotene. In contrast, bell peppers (*Capsicum annuum*) contain a more complex array of carotenoids and, although the mature ripe fruit is red due to the presence of the carotenoids capsanthin and capsorubin, many bell pepper varieties have been developed, which are coloured yellow or orange when mature, and these represent mutants, which lack specific enzymes enabling carotenoid molecules to be interconverted, thereby changing the spectrum of carotenoids within the chromoplasts and the colour of the mature fruit.

Carotenoids have been highlighted as valuable phytonutrients in food and the beneficial effects of eating foods rich in carotenoids have been highlighted. Intake of lycopene has been associated with prevention of cancer and cardiovascular diseases, and intake of the carotenoids in carrots, which are mostly α-carotene and β-carotene, is well known as a source of vitamin A in the diet.

Chromoplasts show extensive heterogeneity in their structure and differ significantly between different tissues, particularly in relation to the type of carotenoids that they accumulate. A scheme of categorisation has been developed to classify chromoplasts according to their structure in considering the nature of the substructures containing pigment that form within them.

Five such categories have been described.

(a) Chromoplasts which are simple in structure and contain globules of pigment accumulated in the stroma.
(b) Chromoplasts, which contain distinct crystals, usually of lycopene or β-carotene.
(c) Chromoplasts, which contain extensive fibrillar or tubular structures.
(d) Membranous chromoplasts, which contain extended concentric membranes.
(e) Reticulo-tubular chromoplasts, which contain a complex network of twisted fibrils throughout the stroma.

A chromoplast of type (a) is shown in Fig. 2.11. The only problem with categorisation of this nature is when chromoplasts exhibit features of more than one class making them difficult to pigeonhole in this way.

Much of the understanding of how chromoplasts develop has come from the detailed examination of the molecular and biochemical changes that occur during ripening of fleshy fruits, especially tomato and bell pepper. During the differentiation of chromoplasts from green-pigmented chloroplasts in these fruits, a controlled breakdown of the chlorophyll and the thylakoid membrane occurs and at the same time, carotenoid biosynthesis increases so that the fruit colour changes from green to red, when fully ripe. Products of early light inducible protein (ELIP) genes appear

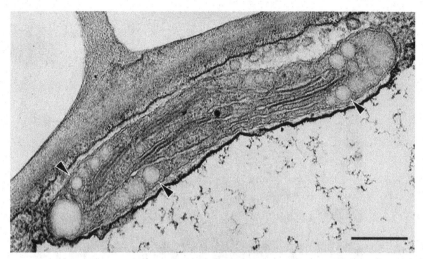

Fig. 2.11. Transmission electron microscope image of a sectioned chromo-plast in a flower petal of marigold (*Tagetes*). The chromoplast contains reduced amounts of thylakoid membrane and large globular bodies, which appear as pale, round structures containing carotenoid pigments (arrowed). Bar = 250 nm. (Image courtesy of http://botit.botany.wisc.edu/images/130/ Plant_Cell/Electron_Micrographs/.)

to be involved in this transition and may play a role in the controlled breakdown of thylakoid membrane in the chloroplast as it undergoes differentiation to become a chromoplast. In mature chromoplasts, there is little thylakoid membrane left and very low levels of chlorophyll. Several other nuclear genes are upregulated during this transition phase. Amongst these are the genes encoding several enzymes involved in carotenoid biosynthesis including phytoene synthase, phytoene desaturase, 1-deoxy-D-xylulose 5-phosphate synthase and a terminal oxidase enzyme, which is associated with phytoene desaturation (see Chapter 7). Genes encoding proteins which are involved in binding the carotenoid molecules into structures are also upregulated, including the protein fibrillin, which is a key component of the fibril structures in which carotenoid molecules are arranged along with lipids within the chromoplast.

Elaioplasts

In addition to starch and coloured pigments, plastids are also capable of storing other types of storage molecules, namely lipid or protein. Plastids

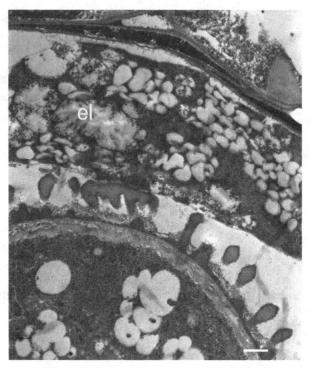

Fig. 2.12. Electron micrograph of a tapetal cell in an anther of *Arabidopsis thaliana*, showing part of an elaioplast (el) containing large numbers of plastoglobuli, which in this image appear as large pale bodies within the plastid body. Scale bar = 200 nm. (Image courtesy of Zoe Wilson and Gema Vizcay-Barrena, Division of Plant and Crop Sciences, University of Nottingham, UK.)

are capable of accumulating lipids in special bodies within the stroma, termed plastoglobules. In most plastids, these are relatively scarce but, in some circumstances, significant numbers of plastoglobules accumulate and, in such situations, the plastids are considered specialised lipid stores and are termed elaioplasts (Fig. 2.12). Elaioplasts occur in a variety of different tissue types throughout plants in which lipid storage is required or where lipids are a better strategy for energy storage than starch. In tapetal cells of the anthers in flowers, elaioplasts accumulate large amounts of neutral esters, which are released when the elaioplast breaks down and the lipids become components of the wall of the pollen grain. In some species of cacti, the plastids in the main body of the cactus accumulate spectacular arrays of plastoglobuli of varying sizes, and in these plastids, lipid is the preferred mode of storage over starch.

Elaioplasts contain several proteins including fibrillin, which forms a coat around the exterior of the plastoglobule preventing them from coalescing together in the plastid and remaining distinct entities. They also contain enzymes involved in the metabolism of iso-prenoid-derived molecules which implies that plastoglobuli have a metabolic role within the plastid rather than just being a site of storage. Originally, plastoglobules were viewed as isolated lipid bodies within the stroma, but more recently it has become clear that they are physically attached to the thylakoid membrane at the edges of granal stacks where the thylakoid membrane is most curved by a half lipid bilayer sheet, which covers part of the plastoglobulus or a group of them. Plastoglobules can also be interconnected by thin neck regions producing groups of plastoglobuli like a string of beads. Since they are intimately connected with the thylakoid membranes in this way, plastoglobules should not only be considered a store of lipid molecules but also of molecules such as carotenoids, plastoquinone and tocopherols, all of which play a role in functionality of the thylakoid membrane and which are fully exchangeable between the thylakoid plastoglobule compartments.

Leucoplasts

Leucoplast is the general name given to the group of plastids which lack pigment and are often called non-green plastids. The amyloplast, elaioplast and proteinoplast could indeed be considered specialised forms of leucoplasts in that they accumulate large amounts of a specific storage molecule but lack pigments. Leucoplasts occur in many plant tissues including stems, seed endosperm, white petals and especially in roots. Indeed, root plastids could be considered a particular type of leucoplast, although clear differences between leucoplasts from different tissues have not been demonstrated. Most leucoplasts contain a little thylakoid membrane and small numbers of plastoglobuli. Although lacking pigments or significant storage materials, leucoplasts play a major role in cellular metabolism (see Chapter 7) particularly in roots. Root leucoplasts differentiate from proplastids in the root apical meristem, lose any thylakoid-like structures and transiently accumulate starch before becoming highly amoeboid in shape (Fig. 2.13). Their populations are relatively low in root cells grown in the dark but, in many species, when roots are grown in the light, some root plastids develop as chloroplasts yielding green roots. This phenomenon appears to be confined normally to those cells

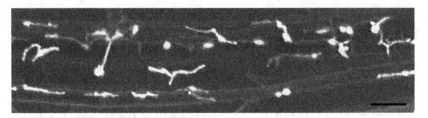

Fig. 2.13. Unpigmented plastids in roots can be visualised by targeting green fluorescent protein (GFP) to them in transgenic *Arabidopsis* seedlings. Fluorescent images, shown here in grey scale, show the root plastids in an individual cell to be highly variable in shape and are maintained as a relatively small population within the cell. Bar = 20 μm.

immediately adjacent to the vasculature in the middle of the root. Root plastids also demonstrate extensive structures called stromules, which are discussed in greater detail in Chapter 8.

Proteinoplasts

In the past, plastids that accumulate significant levels of protein have been termed proteinoplasts or proteoplasts, aleuroplasts and aleurona-plasts although in recent times, such names seem to have been lost from the plastid literature. All photosynthetic plastids contain significant levels of protein in the stroma, primarily that of the carbon dioxide fixation enzyme, ribulose 1,5 bisphosphate carboxylase, usually termed RUBISCO, which can accumulate to significant levels in the stroma and in extreme circumstances can form crystalline arrays of the holoenzyme.

Gerontoplasts

During the senescence of green leaves, there is substantial controlled breakdown of chloroplast components, especially of the proteins of the stroma and thylakoid membrane as well as the thylakoid membrane itself and the chlorophyll it contains. This conversion process gives rise to senescent chloroplasts termed gerontoplasts (Fig. 2.14). The development of gerontoplasts in senescent tissues results in breakdown and loss of the thylakoid membrane network and the accumulation of many plasto-globuli, which themselves contain undigested or indigestible lipophilic molecules, derived from the thylakoid membrane and also contain phytol, which results from chlorophyll breakdown. In some respects,

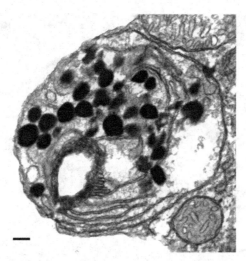

Fig. 2.14. An electron micrograph of a sectioned gerontoplast from a senescent leaf of the grass *Lolium temulentum*. The leaf was senesced artificially on damp filter paper in the dark for 6 days at 20 °C. The resulting gerontoplasts in the leaf mesophyll cells show a loss of the typical oblate spheroid shape of the chloroplasts, partial or complete loss of the thylakoid membrane and granal stacking, and accumulation of many lipid-containing plastoglobuli, which appear as circular black bodies in the stroma. Bar = 200 nm. (Image courtesy of Helen Ougham, IGER, Aberystwyth, Wales.)

gerontoplasts are similar to chromoplasts, both having redifferentiated from chloroplasts, although gerontoplasts do not accumulate newly synthesised pigments.

Etioplasts

A fundamental stimulus for the correct development of chloroplasts in plants is light. In conditions where seeds germinate in the dark or in which leaves develop in the dark, chloroplasts do not undergo correct development and form plastids called etioplasts. See Chapter 6 for a discussion of light-driven chloroplast development. Etioplasts contain limited amounts of thylakoid membrane and often contain starch grains. A major characteristic of their structure is a crystalline array called the prolamellar body (Fig. 2.15). The synthesis of chlorophyll in Angiosperms requires light, particularly in the latter stages; and without light the light-requiring enzyme protochlorophyllide reductase (POR) cannot convert protochlorophyllide into chlorophyllide, the penultimate step in chlorophyll biosynthesis. The prolamellar body is a complex of large amounts of the enzyme

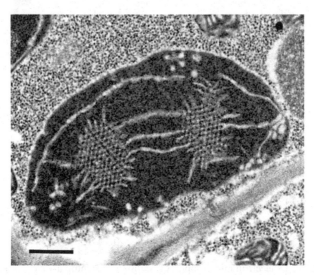

Fig. 2.15. Transmission electron microscope image of a sectioned etioplast from a dark grown seedling. This etioplast contains a little thylakoid membrane and two large crystalline arrays called the prolamellar bodies. Bar = 1 μm. (Image courtesy of Enrique López, Royal Holloway, University of London.)

POR complexed with its substrate protochlorophyllide and lipids. On illumination, the reaction catalysed by POR proceeds and the prolamellar body disperses as the etioplast converts into a chloroplast. In effect, the etioplast structure is primed for rapid conversion to a chloroplast when light appears. The role of etioplasts in nature is minor, since the vast majority of plant development occurs in the light. However, should some plant development occur in a darkened environment, such as a transient cover being placed on growing plants, then etioplasts will form either from proplastids or from the differentiation from existing chloroplasts, which lose their chlorophyll and much of their thylakoid membrane in the process. In the past the etioplast to chloroplast transition has been used extensively in chloroplast development studies, mainly as a system of convenience since seedlings grown in the dark will green up on exposure to light. It is now considered that this system represents a rather artefactual type of chloroplast development and is probably not a good system in which to study authentic chloroplast development.

3

The plastid genome – structure, transcription and translation

The fact that plastids contain their own DNA has been known since the 1960s and has been a central tenet in the construction of theories about plastid evolution and endosymbiosis. The genome of the plastid in higher plants contains approximately 120–160 genes, the products of which function in the plastid in a variety of ways, most importantly in the process of photosynthesis, but also in the translation machinery of the plastid itself. All of the proteins which are encoded by the plastid genome function within the plastid and are not exported. Thus the plastid genome is a distinct 'in-house' genome, which is replicated and transcribed to produce proteins crucial for plastid function. Many of the genes that were present on the genome of the original endosymbiont have migrated into the nuclear genome, leaving behind a residual number of genes which constitute the modern plastid genome that is found today in higher plants. In this chapter we consider how the plastid genome is structured and stored in the plastid, the sequence of the plastid genome and what it encodes and how mRNA molecules resulting from transcription of the plastid genome are translated within the plastid itself.

The architecture of the plastid genome

The DNA in the plastid (ptDNA) has long been considered to exist as a closed, circular molecule of double-stranded DNA of between 120 and 160 kb in size. The evidence for such a circular structure came largely from scanning electron micrographs of isolated ptDNA, which show distinct circular structures, along with shorter linear DNA molecules (Fig. 3.1). In addition, plastid DNA has been analysed by restriction mapping in which isolated DNA is cut into distinct fragments with

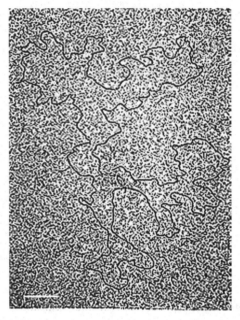

Fig. 3.1. Scanning electron microscopic image of a chloroplast DNA molecule isolated from a soybean chloroplast showing circularity of the molecule. Bar = 1 μm. (From Chu NM, Tewari KK (1982). The arrangement of the ribosomal RNA genes in chloroplasts of *Leguminosae*. *Molecular and General Genetics*, 18, 623–632 with kind permission of Springer Science + Business Media.)

restriction enzymes, and the resulting fragments analysed for their size and sequence. Both of these approaches led to the conclusion that all ptDNA molecules were circular and, as a result, the standard ptDNA maps of gene location and sequence were circular. An explanation for shorter linear molecules observed in preparations was that they represented broken circles of DNA formed during biochemical isolation and thus this theory of ptDNA structure was termed the Broken Circles theory. It was considered that the plastid genome was largely structured like that of bacterial chromosomes and several characteristics are shared between the two.

Large DNA molecules, isolated from chloroplasts which have been carefully lysed within agarose gel, can be separated by pulsed-field gel electrophoresis and, when such gels are stained with the fluorescent DNA stain ethidium bromide, they show clear bands of ptDNA but also reveal that much of the ptDNA remains stuck in the well and does not travel through the gel at all. It was thought originally that this trapped DNA

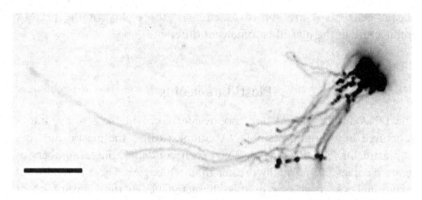

Fig. 3.2. Multigenomic chloroplast DNA structure isolated from meristematic tissue of maize and visualised by staining with the fluorescent DNA binding molecule ethidium bromide. Bar = 10 μm. (From Bendich A (2004). Circular chloroplast chromosomes; the grand illusion. *Plant Cell* 16, 1661–1666. © American Society of Plant Biologists and reprinted with permission.)

was composed of large interlinked circular molecules, which were too big to move through the agarose, but it is now clear that this is not true. Moreover, Bendich and co-workers at the University of Washington, Seattle have shown that the majority of DNA within a plastid is not circular but takes up an array of linear forms, many of them branched due to ongoing replication of the DNA. Consequently, these complex linear arrays of DNA represent structures, which are much bigger than one genome copy, i.e. one complete set of genes encoded by the genome, and are termed multigenomic. Such complexes of linear DNA which take up this architecture can also be seen by modern high-definition imaging (Fig. 3.2). The way in which this linear DNA is replicated is by an interaction of single linear genomes through recombination, which then initiate replication to produce two joined genomes in line. Alternatively, a different interaction between two linear DNA molecules can give rise to two joined genomes in opposite orientations. Also the replication of another molecule can be initiated by transferring the replication fork by recombination. Since replication can be initiated on a linear molecule whilst it is involved in replication in another region, complex arrays of replicating molecules are likely to result. The actual proportion of ptDNA that exists in a circular form is now thought to be small and those molecules which are circular in nature can form a variety of different structures including circles that contain more than one genome copy and groups of interlocking circles called catenanes. Thus, even though plastid gene maps are shown as circular, the reality of the ptDNA architecture in the plastid is somewhat different.

Plastid nucleoids

The DNA in the plastid does not reside naked in the stroma, but is structured in a large protein–DNA complex called the plastid nucleoid or plastid nucleus. Plastid nucleoids each contain variable numbers of genome copies but normally greater than ten copies. There is significant variation between species in this characteristic, in that mature wheat chloroplasts contain 10–30 nucleoids with 70–80 plastid genomes per nucleoid, whereas mature tobacco chloroplasts contain 8–40 nucleoids each containing ten genomes. Numbers of nucleoids within a plastid also vary with plastid type and developmental status but, other than in the proplastid, where there is a single nucleoid in its centre, plastids normally contain several nucleoids, which are associated primarily with the inner plastid envelope membrane (Fig. 3.3). Consequently, plastids routinely contain many copies of their genome and are highly polyploid.

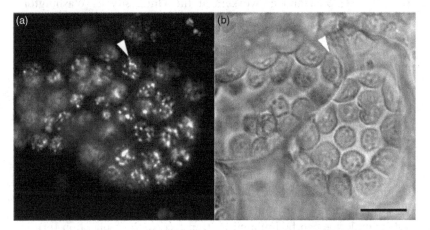

Fig. 3.3. Fluorescence micrographs of plastid nucleoids in mesophyll cells in a young leaf of *Arabidopsis thaliana*. Panel A shows fluorescence of green fluorescent protein (GFP) fused to the plastid nucleoid protein PEND. Panel B shows the same cells viewed with Nomarski optics. The arrowhead in panel A shows nucleoids within chloroplasts and the corresponding position in the Nomarski image is also arrowed. Bar = 10 μm. (From Terasawa K, Sato N (2005). Visualization of plastid nucleoids in situ using the PEND–GFP fusion protein. *Plant and Cell Physiology* 46, 649–660. Reprinted with permission from Oxford University Press.)

The plastid nucleoid contains a variety of DNA-binding proteins, the most abundant of which is the enzyme sulphite reductase. Surprisingly, this enzyme is both a stromal enzyme involved in sulphur metabolism in the plastid (see Chapter 7) and also a major component of the plastid nucleoid, where it induces reversible compaction of the nucleoid thereby enabling or suppressing replication and transcription of the DNA. Another nucleoid protein is PEND (plastid envelope binding protein), which has a DNA binding domain at its N-terminus and a transmembrane region at its C-terminus. PEND is located in the inner plastid envelope membrane and is at least partly responsible for anchoring the nucleoid to the inner envelope membrane. Visualisation of plastid nucleoids in intact plastids within cells can be done by staining isolated plastids with a DNA-binding fluorochrome such as 4',6-diamidino-2-phenylindole (DAPI), which has also been used extensively in fixed tissues and sections. However, non-specific binding to DNA in the nucleus and mitochondria can make interpretation of the resulting images difficult. A more elegant and specific approach is to visualise GFP fused to a nucleoid-localised protein. GFP–PEND protein fusions in transgenic plants reveal elegantly the distribution and morphology of plastid nucleoids (Figs. 3.3, 3.4). Using such an approach, nucleoids can be visualised in different plastid types throughout the plant and are present in non-green plastids as well as chloroplasts. Prior to plastid division, nucleoids replicate, a process which is controlled in part by a DNA gyrase enzyme, which has the ability to change the topology of the plastid DNA within the nucleoid. In chloroplasts undergoing division (see Chapter 8), the nucleoid structure changes from being composed of discrete entities to becoming a network, which eventually separates into the two daughter plastids at the completion of division (Fig. 3.4). After separation of the two daughter plastids, nucleoids revert to being discrete entities once more. HU (also called Hlp) is a DNA-binding protein with histone-like qualities which functions as an architectural protein in the nucleoid and has homologues in the Eubacteria. However, it appears to be only present in nucleoids in algal chloroplasts and not in those of higher plants, suggesting that there are fundamental differences in nucleoid structure between the different groups of photosynthetic organisms. Analysis of the proteome of isolated nucleoid preparations reveals a variety of proteins to be present, some to be expected, such as subunits of DNA and RNA polymerase and topoisomerases and other proteins of unknown function. Exactly how the nucleoid is structured internally with complexes of ptDNA and structural proteins is at present unclear.

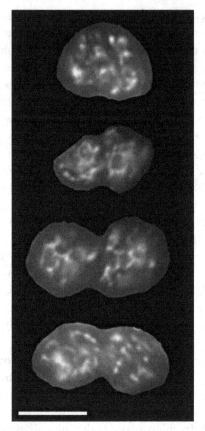

Fig. 3.4. Partition of chloroplast nucleoids during the division of the chloroplast in young cotyledons of *Arabidopsis* and visualised by GFP fluoresence fused to the plastid nucleoid protein PEND. The nucleoids change from being individual structures to forming a network prior to the chloroplast dividing by a central constriction (see Chapter 8 for details of how plastids divide). Bar = 4 μm. (From Terasawa K, Sato N (2005). Visualization of plastid nucleoids in situ using the PEND–GFP fusion protein. *Plant and Cell Physiology* 46, 649–660. Reprinted with permission from Oxford University Press.)

Changes in both nucleoid number and the number of genome copies per nucleoid contribute to changes in the total number of genome copies per plastid during plastid development. Proplastids contain about 20 genome copies, whereas mature chloroplasts contain 100 or more copies of the genome and prior to plastid division this level can increase ten-fold. Since the number of chloroplasts in leaf mesophyll cells is about 100, a mesophyll cell can easily contain 10 000 copies of the plastid genome in its chloroplasts. The number of copies of the genome present in plastids other than chloroplasts is lower, since many of the proteins encoded on the plastid genome are involved in photosynthesis, and this high copy number is required primarily in chloroplasts. There are probably several reasons why plastids maintain such high copy number of their genomes. Firstly, the high copy number facilitates rapid production of proteins constructing ribosomes when rapid plastid development is under way but, perhaps more significantly, high genome copy number, together with high levels of homologous recombination between chloroplast DNA molecules, results in a low mutation rate in chloroplast DNA. Homologous recombination between a mutant gene and a wild-type gene can result in gene conversion, which corrects the mutation. Without such a mechanism, the asexual replication of plastid DNA would quickly lead to a highly mutant and heterogeneous genome, which does not happen in plastids, where the genome is highly conserved and relatively stable.

Plastid genome structure

Since the first plastid genome sequences were determined for tobacco and the liverwort *Marchantia* in 1986, many plant species have had their plastid genomes completely sequenced, representing all the major lineages of plant evolution. Plastid genome sequences for over 100 species are available at http://www.bch.umontreal.ca/ogmp/projects/other/cp_list. html. The plastid genome sequences of land plants show them to be relatively similar and conserved in sequence, but in algae there is much greater variation in genome organisation and coding capacity (Table 3.1). Several general characteristics of higher plant plastid genomes are clear. They contain a high proportion of the bases adenine and thymidine (A and T) compared with the bases cytidine and guanidine (C and G), with 60%–70% of bases being AT, and in non-coding regions this proportion can reach 80%. In coding regions, there is a propensity for codons with A or T in the third position to be used where several different codons can encode the same amino acid. A major structural characteristic of the ptDNA sequence is the presence of two inverted repeat sequences, which are identical in sequence but inverted in orientation (Fig. 3.5) such that they delimitate two regions between them that make up the rest of the genome; a large single copy region (LSC) and a small single copy region (SSC). Thus, genes present in the inverted repeat sequences, which are about 25 kb in size, are present twice in the genome, whereas all other

Table 3.1. *Characteristics of plastid genomes of various species of higher and lower plants derived from the plastid genome database at http://chloroplast.cbio.psu.edu/cgi-bin/organism.cgi*

	Genome size (kb)	Total genes	Protein coding genes	tRNA genes	rRNA genes
ANGIOSPERMS					
Amborella trichopoda	162 686	133	86	36	8
Acorus calamus	153 821	135	85	38	8
Arabidopsis thaliana	154 478	117	87	37	8
Atropa belladonna	156 687	133	87	37	8
Calycanthus floridus	153 337	137	88	37	8
Citrus sinensis	160 129	142	89	45	8
Cucumis sativus	155 293	130	89	38	8
Daucus carota	155 911	138	85	43	8
Drimys granadensis	160 604	139	87	44	8
Epifagus virginiana	70 028	56	26	28	8
Eucalyptus globulus	160 286	135	90	37	8
Glycine max	152 218	128	83	37	8
Gossypium hirsutum	160 301	131	85	37	8
Helianthus annuus	151 104	141	87	44	8
Jasminum nudiflorum	165 121	133	87	38	8
Liriodendron tulipifera	159 886	131	86	37	8
Lactuca sativa	152 765	128	84	37	7
Lotus corniculatus	150 519	127	82	37	8
Lycopersicon esculentum	155 461	133	86	37	8
Morus indica	158 484	130	84	37	8
Nandina domestica	156 599	136	87	37	8
Nicotiana sylvestris	155 941	163	111	37	8
Nicotiana tabacum	155 943	148	102	37	8
Nicotiana tomentosiformis	155 745	162	109	37	8
Nymphaea alba	159 930	138	89	37	8
Oenothera elata	163 935	163	118	38	8
Oryza nivara	134 494	164	119	38	8
Oryza sativa	134 525	157	108	41	8
Panax ginseng	156 318	135	87	38	8
Pelargonium x hortorum	217 942	220	134	40	10
Phalaenopsis aphrodite	148 964	151	97	37	7

Table 3.1. (cont.)

	Genome size (kb)	Total genes	Protein coding genes	tRNA genes	rRNA genes
Piper cenocladum	160 624	132	88	37	8
Platanus occidentalis	161 791	136	85	37	8
Populus alba	156 505	129	83	37	8
Saccharum hybrid cultivar	141 182	144	97	39	8
Saccharum officinarum	141 182	165	117	38	8
Solanum bulbocastanum	155 371	142	87	45	8
Solanum tuberosum	155 312	131	81	37	8
Spinacia oleracea	150 725	140	101	37	8
Triticum aestivum	134 545	134	83	42	8
Vitis vinifera	160 928	140	86	45	8
Zea mays	140 384	160	111	39	8
GYMNOSPERMS					
Pinus koraiensis	116 866	301	164	37	4
Pinus thunbergii	119 707	242	159	37	4
BRYOPHYTES					
Physcomitrella patens	122 890	130	85	37	8
Anthoceros formosae	161 162	136	91	37	8
Huperzia lucidula	154 373	131	87	33	8
Marchantia polymorpha	121 024	135	89	37	8
MONILOPHYTES					
Adiantum capillus-veneris	150 568	130	87	35	8
Psilotum nudum	138 829	150	101	41	8
ALGAE					
Chaetosphaeridium globosum	131 183	141	98	37	6
Chara vulgaris	184 933	148	105	37	6
Chlamydomonas reinhardtii	203 828	99	69	29	10
Chlorella vulgaris	150 613	210	174	33	3
Cyanidioschyzon merolae	149 987	241	207	31	3
Cyanidium caldarium	164 921	230	197	30	3
Cyanophora paradoxa	135 599	191	149	36	6
Emiliania huxleyi	105 309	155	119	30	6
Gracilaria tenuistipitata	183 883	238	203	29	3
Guillardia theta	121 524	183	147	30	6

Table 3.1. (cont.)

	Genome size (kb)	Total genes	Protein coding genes	tRNA genes	rRNA genes
Helicosporidium sp.	37 454	54	26	25	3
Mesostigma viride	118 360	148	105	37	6
Nephroselmis olivacea	200 799	199	155	38	6
Odontella sinensis	119 704	171	140	29	6
Oltmannsiellopsis viridis	151 933	127	93	28	6
Ostreococcus tauri	71 666	94	61	27	6
Porphyra purpurea	191 028	254	209	37	6
Porphyra yezoensis	191 952	264	209	49	6
Pseudendoclonium akinetum	195 867	141	105	31	6
Scenedesmus obliquus	161 452	114	78	30	6
Staurastrum punctulatum	157 089	138	103	32	3
Zygnema circumcarinatum	165 372	140	103	34	3

genes are present only once. Exactly why these inverted repeats are present in the plastid genome and how they function is unclear, since they vary in size between different species and, in some higher plants such as members of the *Leguminosae* and several algal species, they are absent (Fig. 3.6). One function of the inverted repeat may be to increase copy number of important genes, since genes encoding rRNA destined for the plastid's own ribosomes are present within the inverted repeat. In addition, the inverted repeat may contribute stability to the plastid molecule, although such considerations have really only been used when considering the ptDNA as a circular molecule. Recombination between the homologous sequences of the two inverted repeat gives rise to two different isoforms of ptDNA, which differ in the relative orientations of the ptDNA, and the directionality of the LSC in relation to the SSC.

The gene content of Angiosperm plastid genomes varies between 120 and 165 genes in a genome between 120 and 165 kb in size (Table 3.1) but plastid genomes from organisms earlier in the evolutionary progression, through algae to lower plants to higher plants contain some genes present on the plastid genome, which in higher plants, are now present in the nuclear genome. This is exemplified by the fact that the Cyanobacterium *Synechocystis*, which is a present-day relative of the original prokaryotic

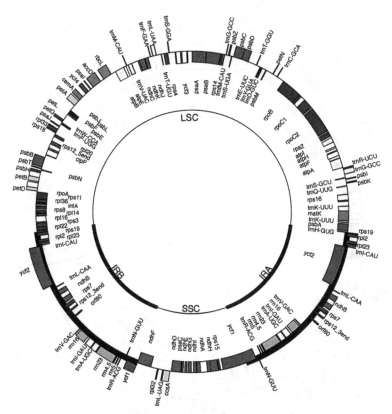

Fig. 3.5. Gene map of the plastid genome of carrot *(Daucus carota)*. The complete genome is 155 911 base pairs of DNA in size and genes are encoded on either of the two DNA strands, shown as on the inside or outside of the circle. Gene names correspond to those listed in Table 3.2, except for those labelled Open Reading Frame (ORF), which define genes of unknown function. The inverted repeat regions are shown on the inner circle as thick black lines named IRA and IRB. Genes within the inverted repeat are duplicated and thus have two copies per genome. The inverted repeat defines two single copy regions of differing sizes, the small single copy (SSC) region and the large single copy region (LSC). (Modified from Ruhlman T, Lee S-B, Jansen RK *et al.* (2006). Complete plastid genome sequence of *Daucus carota*: implications for biotechnology and phylogeny of angiosperms. *BMC Genomics* 7, 222.)

endosymbiont that gave rise to the modern plastid, has a genome containing 3000 genes. Consequently, the genes encoded by the plastid genome in modern-day photosynthetic multicellular organisms only encode a small proportion of the total proteome of the plastid, and around 95% of those proteins present in the mature plastid are encoded

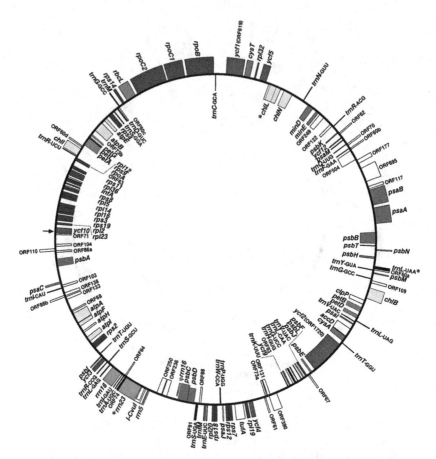

Fig. 3.6. Gene map of the plastid genome from the Green alga *Chlorella vulgaris*. The complete genome is 150 613 base pairs of DNA in size. Genes are encoded on either of the two DNA strands, shown as on the inside or outside of the circle. In contrast to the genome map of carrot (Fig. 3.5), this algal genome lacks the inverted repeat sequences and contains some genes absent from higher plant plastid genomes such as carrot. For example, the plastid division genes *minD* and *minE* are present in the *Chlorella* plastid genome, but in higher plants these genes have migrated to the nuclear genome. (From Wakasugi T, Nagai T, Kapoor M *et al.* (1997). Complete nucleotide sequence of the chloroplast genome from the green alga *Chlorella vulgaris*: the existence of genes possibly involved in chloroplast division. *Proceedings of the National Academy of Sciences*, USA 94, 5967–5972. © 1997, National Academy of Sciences, USA.)

by nuclear genes and imported as proteins that have been translated in the cytoplasm.

The genes encoded by the plastid genome fall into three major groups: genes encoding proteins involved in photosynthesis, genes encoding genetic system proteins within the plastid and a heterogeneous third group of various genes, some of unknown function (Table 3.2). Around 46 genes encode photosynthetic proteins, which are part of the large complexes that are associated with the thylakoid membrane and facilitate light energy capture and photosynthetic electron transport leading to the production of high energy storage molecules, ATP and NADPH (see Chapter 4). Within these complexes, those proteins encoded on the plastid genome form only a subset of the total in each complex, the rest being encoded by nuclear genes and the proteins imported. For instance, photosystem II complex contains about 24 proteins but only 15 are encoded by the plastid genome. The plastid genome also encodes seven proteins functional in photosystem I, six proteins involved in the cytochrome b_6f complex and six proteins which form part of the ATP synthase complex, which is involved in making ATP on the thylakoid membrane. A key protein encoded on the plastid genome is the large subunit (LSU) of the carbon dioxide fixing enzyme, ribulose 1,5 bisphosphate carboxylase/oxygenase (RUBISCO), which forms a holoenzyme in the plastid consisting of eight large subunits and eight small subunits (SSU), the latter being encoded in the nucleus and imported (see Chapter 4). All of the thylakoid photosynthetic complexes and the RUBISCO enzyme, therefore, are genetic hybrids composed of proteins encoded in two distinct genetic compartments: the nucleus and the plastid. The efficient production of such complexes, therefore, requires tight coordination and signalling between the plastid and the nucleus (see Chapter 6).

The majority of the genes in the plastid genome encode proteins involved in genetic systems; that is transcription, RNA processing, translation and protein degradation (Table 3.2). Included in this group are 21 genes and four rRNAs, which are involved in the construction of ribosomes within the plastid, which facilitate translation of mRNA molecules derived by transcription of the plastid genome. The remaining proteins in the plastid ribosomes are nuclear-encoded. All of the tRNAs required for the translation of plastid mRNA molecules representing all 20 amino acids are encoded in the plastid genome. Thus, even in terms of making its own proteins within itself, the plastid is beholden to the nucleus to enable construction of fully functional ribosomes. The machinery involved in plastid transcription is also a genetic hybrid from the two genomes.

Table 3.2. *Genes and conserved open reading frames (ycf) encoded by the plastid genome in higher plants*

Gene	Gene product	Functions and remarks
Photosynthetic system genes		
psaA	A subunit of PSI	reaction centre subunit, essential for PSI function
psaB	B subunit of PSI	reaction centre subunit, essential for PSI function
psaC	C subunit of PSI	essential co-factor-binding subunit
psaI	I subunit of PSI	small subunit, not essential for PSI function
psaJ	J subunit of PSI	small subunit, not essential for PSI function
ycf3	Ycf3 protein	essential PSI assembly factor, contains three tetratricopeptide (TPR) repeats
ycf4	Ycf4 protein	essential PSI assembly factor
psbA	D1 protein of PSII	reaction centre, also termed herbicide-binding protein, essential for PSII function
psbB	CP47 subunit of PSII	inner antenna protein, essential for PSII function
psbC	CP43 subunit of PSII	inner antenna protein, essential for PSII function
psbD	D2 protein of PSII	reaction centre, essential for PSII function
psbE	cytochrome b_{559}	essential for PSII assembly/stability/function, protection of PSII against photoinhibition
psbF	cytochrome b_{559}	essential for PSII assembly/stability/function, protection of PSII against photoinhibition
psbH	H subunit of PSII	small subunit associated with CP47, involved in PSII assembly, stabilisation and photoprotection
psbI	I subunit of PSII	small subunit, involved in stabilisation of PSII dimers and PSII-LHCII supercomplexes
psbJ	J subunit of PSII	small subunit, involved in assembly of the water-splitting complex and intra-complex electron transfer
psbK	K subunit of PSII	small subunit associated with CP43, presumably involved in PSII assembly/stability

psbL — L subunit of PSII, small subunit, involved in PSII dimerisation and PSII-LHCII supercomplex formation, required for assembly of the water-splitting complex

psbM — M subunit of PSII, small subunit, function unknown
psbN — N subunit of PSII, function unknown, assignment as PSII subunit uncertain
psbT — T subunit of PSII, small subunit, involved in repair of photodamaged PSII reaction centres

psbZ — Z subunit of PSII, small subunit, couples the light-harvesting complex protein CP26 to PSII

petA — cytochrome *f*, core subunit of cyt b_6f complex, essential for cyt b_6f function
petB — cytochrome b_6, core subunit of cyt b_6f complex, essential for cyt b_6f function
petD — subunit IV of cyt b_6f, essential for cyt b_6f function
petG — G subunit of cyt b_6f, small subunit, essential for cyt b_6f assembly/stability in *Chlamydomonas*
petL — L subunit of cyt b_6f, small subunit, not essential for cyt b_6f function, involved in complex stabilisation

petN — N subunit of cyt b_6f, small subunit, essential for cyt b_6f assembly/stability
atpA — ATP synthase α subunit, CF_1, nucleotide binding site
atpB — ATP synthase β subunit, CF_1, catalytic binding site
atpE — ATP synthase ε subunit, CF_1, regulation of CF_1/CF_0 activation, required for proton gating
atpF — ATP synthase b-subunit, CF_0, binding of CF_1
atpH — ATP synthase c-subunit, CF_0, proton translocation
atpI — ATP synthase a-subunit, CF_0, proton translocation
ndhA — A subunit of NAD(P)H dehydrogenase, cyclic electron transfer, chlororespiration

ndhB — B subunit of NAD(P)H dehydrogenase, cyclic electron transfer, chlororespiration

ndhC — C subunit of NAD(P)H dehydrogenase, cyclic electron transfer, chlororespiration

ndhD — D subunit of NAD(P)H dehydrogenase, cyclic electron transfer, chlororespiration

Table 3.2. (cont.)

Gene	Gene product	Functions and remarks
ndhE	E subunit of NAD(P)H dehydrogenase	cyclic electron transfer, chlororespiration
ndhF	F subunit of NAD(P)H dehydrogenase	cyclic electron transfer, chlororespiration
ndhG	G subunit of NAD(P)H dehydrogenase	cyclic electron transfer, chlororespiration
ndhH	H subunit of NAD(P)H dehydrogenase	cyclic electron transfer, chlororespiration
ndhI	I subunit of NAD(P)H dehydrogenase	cyclic electron transfer, chlororespiration
ndhJ	J subunit of NAD(P)H dehydrogenase	cyclic electron transfer, chlororespiration
ndhK	K subunit of NAD(P)H dehydrogenase	cyclic electron transfer, chlororespiration
rbcL	Rubisco large subunit	CO_2 fixation
Genetic system genes		
rpoA	RNA polymerase subunit	transcription, E.coli-like plastid RNA polymerase (PEP)
rpoB	RNA polymerase subunit	transcription, E.coli-like plastid RNA polymerase (PEP)
rpoC1	RNA polymerase subunit	transcription, E.coli-like plastid RNA polymerase (PEP)
rpoC2	RNA polymerase subunit	transcription, E.coli-like plastid RNA polymerase (PEP)
matK	intron maturase	splicing factor for group II introns
rrn16	16S ribosomal RNA	translation, small ribosomal subunit
rrn23	23S ribosomal RNA	translation, large ribosomal subunit
rrn5	23S ribosomal RNA	translation, large ribosomal subunit
rrn4.5	23S ribosomal RNA	translation, large ribosomal subunit
trnA-UGC	tRNA-Alanine(UGC)	translation
trnC-GCA	tRNA-Cysteine(GCA)	translation

trnD-GUC	tRNA-Aspartate(GUC)	translation
trnE-UUC	tRNA-Glutamate(UUC)	translation, tetrapyrrole biosynthesis
trnF-GAA	tRNA-Phenylalanine (GAA)	translation
trnG-GCC	tRNA-Glycine(GCC)	translation
trnG-UCC	tRNA-Glycine(UCC)	translation
trnH-GUG	tRNA-Histidine(GUG)	translation
trnI-CAU	tRNA-Isoleucine(CAU)	translation
trnI-GAU	tRNA-Isoleucine(GAU)	translation
trnK-UUU	tRNA-Lysine(UUU)	translation
trnL-CAA	tRNA-Leucine(CAA)	translation
trnL-UAA	tRNA-Leucine(UAA)	translation
trnL-UAG	tRNA-Leucine(UAG)	translation
trnM-CAU	tRNA-Methionine(CAU)	translation
trnfM-CAU	tRNA-N-Formyl- methionine (CAU)	translation initiation
trnN-GUU	tRNA-Asparagine(GUU)	translation
trnP-UGG	tRNA-Proline(UGG)	translation
trnQ-UUG	tRNA-Glutamine(UUG)	translation
trnR-ACG	tRNA-Arginine(ACG)	translation
trnR-UCU	tRNA-Arginine(UCU)	translation
trnS-GCU	tRNA-Serine(GCU)	translation
trnS-GGA	tRNA-Serine(GGA)	translation
trnS-UGA	tRNA-Serine(UGA)	translation
trnT-GGU	tRNA-Threonine(GGU)	translation
trnT-UGU	tRNA-Threonine(UGU)	translation
trnV-GAC	tRNA-Valine(GAC)	translation
trnV-UAC	tRNA-Valine(UAC)	translation
trnW-CCA	tRNA-Tryptophan(CCA)	translation
trnY-GUA	tRNA-Tyrosine(GUA)	translation

Table 3.2. (cont.)

Gene	Gene product	Functions and remarks
rps2	ribosomal protein S2	translation, small ribosomal subunit
rps3	ribosomal protein S3	translation, small ribosomal subunit
rps4	ribosomal protein S4	translation, small ribosomal subunit
rps7	ribosomal protein S7	translation, small ribosomal subunit
rps8	ribosomal protein S8	translation, small ribosomal subunit
rps11	ribosomal protein S11	translation, small ribosomal subunit
rps12	ribosomal protein S12	translation, small ribosomal subunit
rps14	ribosomal protein S14	translation, small ribosomal subunit
rps15	ribosomal protein S15	translation, small ribosomal subunit
rps16	ribosomal protein S16	translation, small ribosomal subunit
rps18	ribosomal protein S18	translation, small ribosomal subunit
rps19	ribosomal protein S19	translation, small ribosomal subunit
rpl2	ribosomal protein L2	translation, large ribosomal subunit
rpl14	ribosomal protein L14	translation, large ribosomal subunit
rpl16	ribosomal protein L16	translation, large ribosomal subunit
rpl20	ribosomal protein L20	translation, large ribosomal subunit
rpl22	ribosomal protein L22	translation, large ribosomal subunit
rpl23	ribosomal protein L23	translation, large ribosomal subunit
rpl32	ribosomal protein L32	translation, large ribosomal subunit
rpl33	ribosomal protein L33	translation, large ribosomal subunit
rpl36	ribosomal protein L36	translation, large ribosomal subunit
infA	translation initiation	inactive pseudogene or gene lost

Other genes

clpP	catalytic subunit of the protease Clp	ATP-dependent protein degradation, essential for cell survival
accD	acetyl-CoA carboxylase subunit A of the system II	subunit fatty acid biosynthesis, essential for cell survival
ycf5 / ccsA	complex for c-type cytochrome biogenesis	required for haem attachment to chloroplast c-type cytochromes
ycf10	inner envelope protein	presumably involved in the uptake of inorganic carbon
ycf1	putative Ycf1 protein	essential gene, function unknown
ycf2	putative Ycf2 protein	essential gene, function unknown, contains a putative nucleotide-binding domain
ycf15	unknown ORF	unclear functional significance
sprA	small RNA	function unknown

Source: Adapted from Bock R (2007). Structure, function and inheritance of plastid genomes. In (Bock R ed.) *Chloroplasts. Topics in Current Genetics* 19 with kind permission of Springer Science + Business Media.

The four subunits of an RNA polymerase (PEP) are encoded in the plastid genome. In contrast, a second RNA polymerase (NEP) found in the plastid is entirely encoded in the nucleus (see next section).

The third group of genes encoded by the plastid genome represent a mixed bag including some open reading frames, which are conserved between plastid genomes in different species and are considered to be genuine genes. In such cases, they have been termed *ycf* genes (hypothetical *c*hloroplast reading *f*rame). Other non-conserved open reading frames between species are likely to lack any significant function within the plastid. Three other proteins encoded within this third group include the protein required for haem attachment to cytochrome molecules within the plastid (*ccsA*), a gene which encodes a subunit of acetyl-CoA-carboxylase (*accD*), which is central to fatty acid biosynthesis within the plastid, and a subunit of a protease involved in protein degradation within the plastid (*ccpP*).

Although the plastid genomes of Angiosperms vary in size to some extent between different species, variation in gene content is largely a result of differences in the size of the inverted repeat where those genes present are included twice in the total gene count (Table 3.1), and in general the plastid genome is highly conserved. However, some species have significantly different-sized plastid genomes. Pelargonium has a particular large genome containing over 200 genes and two pine species with sequenced genomes have small-sized genomes since they have very small inverted repeats, but which are very gene rich and contain a large number of genes (Table 3.1).

A particular class of plants in which the plastid genome is reduced significantly in size is parasitic plants. The plastid genome of the root holoparasite *Epifagus virginiana* is less than half the size of plastid genomes of properly photosynthetic Angiosperms and contains only 26 protein coding genes (Table 3.1), which are mostly of the genetic system group. Thus in this parasitic plant, which exists heterotrophically as a result of its parasitism, all of the photosynthetic genes normally encoded on the plastid genome have been lost. Exactly why such a plant bothers to retain many of the plastid genetic system genes on its reduced genome is unclear, especially since not all of the conventional genetic system genes are present, yet the remaining plastid genes of *Epifagus* are expressed normally. Other parasitic plant species such as those in the genus *Cuscuta*, which carry out varying degrees of photosynthesis according to the nature of their parasitism, have variable plastid genomes, which still contain some photosynthetic genes. Such a group appear to be on the evolutionary pathway of plastid genome loss as heterotrophy evolves.

Compared with the Angiosperm plastid genomes, those of algae are much more variable in size and gene content and many different changes have occurred during different strands of algal evolution. Some species such as *Chlorella vulgaris* lack the inverted repeat (Fig. 3.6) and some Red and Green algae species have a directed repeat rather than an inverted repeat. The most extreme change has been in some Dinoflagellates, where the genome is made up of many minicircles of DNA, 2–3 kb in size, each of which contain 1–3 genes and which collectively add up to the whole genome, which is significantly reduced in total gene content. Gene density also varies enormously between different algal genomes. For instance, the genome of the Cryptophyte alga *Guillardia theta* has 183 genes in a genome of 122 kb, whereas the Green alga *Chlamydomonas reinhardtii* has a large genome of 204 kb but only has 99 genes, which is mostly as a result of short repeated DNA sequences in the intergenic regions. Some algae possess genes on their plastid genome, which in the more evolutionary advanced higher plants have moved to the nucleus. A good example of this are two *Min* genes, which encode proteins involved in plastid division, which are present on the plastid genome of *Chlorella vulgaris* (Fig. 3.6) but are absent in the plastid genome of higher plants (Fig. 3.5, Table 3.2), where they are now present in the nucleus.

A particularly novel example of a plastid genome occurs in members of the group of animal parasites: the *Apicomplexa*. Famous amongst these are *Plasmodium falciparum*, which causes malaria and *Toxoplasma gondii*, which causes toxoplasmosis. These organisms contain an apicoplast, which is an essential plastid-like organelle and appears to have evolved by a secondary endosymbiotic event (see Chapter 1). The apicoplast contains a small genome encoding about 60 genes, mostly involved in encoding genetic system genes and which are actively transcribed and translated in the apicoplast. Exactly why this genome has been retained actively in these parasites is unclear, but the most likely reason is that the apicoplast is a site of essential cellular metabolism and remains functional in these parasites for that reason. Even so, virtually all of the metabolic reagents and enzymes need to be imported into the relic plastid in order to function.

Transcription of the plastid genome

The plastid genome is transcribed using three different RNA polymerase–enzyme complexes. One RNA polymerase complex is encoded in the plastid genome by four genes encoding its four core subunits, RpoA, RpoB, RpoC1 and RpoC2 and is termed the plastid-encoded polymerase (PEP). These four genes are transcribed from the plastid genome and the proteins are translated by using plastid ribosomes inside the plastid. This PEP enzyme complex is similar to that found in *Escherichia coli* and uses a eubacterial-type transcription machinery to facilitate transcription of certain genes encoded on the plastid genome. Although the four plastid-encoded proteins form the core, polymerase isolated fractions from plastids with PEP activity suggest that the functional enzyme complex contains other components including RNA binding proteins and RNA processing proteins, both of which are nuclear-encoded. The PEP complex initiates transcription of plastid-encoded genes by recognising conserved -35 and -10-like promoter sequences, but an essential component required for this process is a sigma subunit, which determines promoter specificity and is nuclear-encoded. In fact, sigma factors add an extra level of complexity and control to the activity of the PEP complex in the plastid since they are nuclear-encoded and thus enable the nucleus to play a role in controlling PEP activity in the distant plastid organelle. There are several nuclear genes encoding sigma factor proteins and it remains to be discovered exactly how they function, either combinatorially or individually to bring about PEP-mediated transcription of plastid DNA at different stages of development and in different tissues. Of the six sigma factor genes encoded by the *Arabidopsis* genome, only two, *AtSig2* and *AtSig5*, appear crucially important in plastid function, since mutating either gene results in impaired plastid development. An important characteristic of sigma factor genes is the fact that their expression is positively regulated by light and the expression of the *AtSig5* is induced specifically by blue light. Thus, a mechanism exists for the control of PEP activity in the plastid by light via the light-controlled expression of sigma factors genes in the nucleus (see Chapter 6).

The fact that the transcription of plastid-encoded genes can still occur in plants where plastid ribosomes are deficient or where a PEP subunit has been knocked out artificially by genetic means, suggested that there is a second type of RNA polymerase at work in the plastid. This second RNA polymerase was found to be a single subunit molecule, which shows homology to polymerases from bacteriophage and the mitochondria of yeast and is a nuclear-encoded polymerase, termed NEP. Surprisingly, there are three nuclear genes encoding NEP, which are imported into different cellular organelles; one NEP is specific for the plastid, another

NEP is specific for the mitochondria and a third NEP appears to be targeted to both organelles. Thus the plastid contains two NEP-type RNA polymerases. NEP recognises distinct types of promoters associated with plastid-encoded genes, the basic characteristic of which is a YRTA motif.

There is a distinct sequential hierarchy to the mode of action of these PEP and NEP RNA polymerases during early development of the chloroplast from the proplastid. Initially, the NEP gene is expressed in the nucleus and NEP imported into the proplastid, where it transcribes primarily the genetic system genes on the plastid genome, including the four genes encoding the PEP–enzyme complex. Once PEP proteins have been translated on the plastid ribosomes and the correct sigma factors have been imported, PEP becomes active and then transcribes the photosynthetic group of genes on the chloroplast genome. In addition, NEP continues to be active in mature chloroplasts and several of the photosynthetic plastid-encoded genes can be transcribed by both NEP and PEP. In non-green plastids, such as those in roots, NEP is the main polymerase functional in the plastid since activity of PEP is governed by the presence of sigma factors in the plastid which are required for PEP activity, the expression of which is light-stimulated.

There are additional levels of complexity in this system. Once PEP is active and transcribing plastid genes, it appears that one product of PEP transcription, glutamyl-tRNA, which is also a precursor of chlorophyll biosynthesis, binds to NEP molecules in the plastid repressing their activity, such that PEP then becomes the dominant polymerase as the plastid develops. From a concentration point of view, PEP will be more abundant in the mature chloroplast simply because the copy number of its genes are significantly greater than that of the single NEP genes in the nucleus.

Plastid RNA processing

The actual array of transcripts produced by NEP and PEP transcription activity on the plastid genome is complex since much of the transcription of the plastid genome is polycistronic; that is, several genes which are sequentially placed on the genome are transcribed together into one primary mRNA molecule. Such polycistronic transcripts are then cleaved by endonuclease enzymes in a complex pattern giving rise to a variety of overlapping transcripts, including monocistronic mRNAs representing a single gene. In addition, transcription can be initiated at a variety of alternative promoters within a polycistronic region, which also produces a variety of overlapping transcripts. Consequently, RNA processing is an important regulatory step in plastid gene expression. In higher plant plastid DNA, most genes are transcribed via a polycistronic transcription, although a few genes, such as the large subunit of RUBISCO (*rbcL*) and most of the genes encoding tRNAs, are transcribed individually as a monocistronic transcription unit (Table 3.3). In many cases, the polycistronic transcription units contain genes with a related function within the plastid. For example, four different *rDNA* genes encoding ribosomal RNAs destined for plastid ribosome construction are transcribed polycistronically, as are ten of the plastid ribosomal protein genes, which are co-transcribed in a cistron containing 12 genes (Table 3.3). Other polycistronic transcription units contain genes with unrelated functions (Table 3.3). Processing of a primary polycistronic mRNA molecule is complex and can yield a large number of individual molecules, each of which appears capable of being translated into functional protein. For instance, one polycistronic transcription unit contains four genes encoding components of both photosystems, *psbB–psbH–petB–petD*. The tetracistronic primary RNA molecule, which results from transcription of these four genes together, subsequently undergoes a complex set of processing reactions, which gives rise to around 20 intermediate mRNA molecules, including the monocistronic mRNAs for the four individual genes. However, this is further complexed by the fact that, in different species, processing occurs to differing extents, such that, in spinach, *psbH* and *psbB* are translated from a discistronic molecule containing both genes. Although all this processing occurs, the majority of the intermediates and final processed mRNA molecules can be translated efficiently into proteins by the plastid ribosomes.

Even though plastids have a prokaryotic evolutionary history, many plastid genes contain introns, which require splicing out from the primary RNA transcript prior to translation. In the tobacco plastid genome, there are 18 genes containing a total of 20 introns, classified as types I, II or III. In some cases, this splicing occurs spontaneously, since the mRNA molecule possesses splicing ability itself and splices out its own intron, by a process termed self-splicing. In other cases, nuclear-encoded proteins are required in order for splicing to occur within the plastid. For example, splicing of the intron in the *atpF* gene, which encodes a subunit of the ATPase synthase complex on the thylakoid

Table 3.3. *A generalised listing of the monocistronic and polycistronic transcription units in the plastid DNA of higher plants. Monocistronic transcription units transcribe single genes and the mRNA is translated into protein conventionally whereas polycistronic transcription units result in transcription of several genes into a single primary RNA transcript which is then spliced in a variety of ways to yield translatable mRNA molecules.*

MONOCISTRONIC TRANSCRIPTION UNITS
ndhF
psbA
psbM
psbN
rbcL
most of the *tRNA* genes

POLYCISTRONIC TRANSCRIPTION UNITS
Genes with related function
atpB-atpE
3′ rps-12-rps7
psbE-psbF-psbL-psbJ
psbD-psbC-orf62 (in dicots)
psbK-psbI-psbD-psbC-orf62-trnG (in monocots)
ndhC-ndhK-ndhJ
16SrDNA-trnI-trnA-23SrDNA-4.5SrDNA-5SrDNA
rpoB-rpoC1-rpoC2
rpl23-rpl12-rps9-rpl22-rps3-rpl16-rpl14-rps8-infA-rpl36-rps11-rpoA
trnE-trnY-trnD

Genes with unrelated function
clpP-5′rps12-rpl20
orf31-petG-psaJ-rpl33-rps18
psaA-psaB-rps14
psaC-ndhD
psbB-psbH-petB-petD
psbK-psbI-trnG
ndhA-ndhI-ndhG-ndhE-psaC
rpl32-sprA
rps2-atpI-atpH-atpF-atpA

Source: Adapted from Sugita M, Sugiura M (1996) *Plant Molecular Biology* 32, 315–326.

membrane, requires a nuclear-encoded RNA binding protein, CRS1 which probably enables the pre-mRNA molecule to undergo self-splicing itself. In the most extreme example of splicing, a chloroplast protein is encoded by regions of DNA in different parts of the plastid genome, each of which is transcribed and the individual mRNA molecules are joined together in a process termed trans-splicing. An example of trans-splicing is shown by the *rps12* gene, encoding plastid ribosomal subunit 12, which is composed of three exons. Exons 2 and 3 are transcribed together in one unit and exon 1 is transcribed separately from a gene in a distant part of the genome. In fact, exon 1 is part of a tricistronic unit, with two other genes, and exons 2 and 3 are separated by an intron sequence and co-transcribed with another gene in a dicistronic unit (Table 3.3). Once processed, these two mRNA molecules are then spliced together before translation of the complete rps12 protein can occur.

mRNA molecules derived from transcription of the plastid genome also undergo a process of RNA editing, in which specific nucleotides are interchanged, resulting in a base sequence in the mRNA molecule, which does not match that of the DNA template sequence. In chloroplasts, most editing results in changes of specific bases, mostly conversion of C to U, which normally results in a change in the amino acid coding potential of the mRNA. In general, the RNA editing process restores codons, which are highly conserved and thus are likely to be crucial in the function of the resulting protein. Experiments where non-edited mRNAs are translated in the plastid result in proteins that function poorly, if at all. In the instance of the *ndhD* transcript, editing results in the creation of the initiation codon for translation, such that failure to edit correctly leads to a loss of ndhD protein synthesis. Although RNA editing is relatively extensive with around 36 specific editing sites known in plastids, little else is known of the precise biochemistry by which RNA editing in plastids occurs. A complex carrying out this process in the plastid, termed the editosome, has been postulated but the nature of its components and how it functions remain unclear.

A further important factor in the molecular biology of plastid gene transcription is that of RNA stability. There is much variation between different mRNA molecules and transcription units in their levels of stability, essentially reflected in their time spent in the stroma prior to degradation, as well as significant changes in stability of a given mRNA molecule with the developmental state of the plastid and environmental conditions. Most transcription units have a short inverted repeat sequence at their 3′ untranslated region, which acts as a signal for RNA processing, and such sequences are crucial for the stability of these transcriptional units. When they are removed experimentally, the RNA molecules are rapidly degraded. A large number of nuclear-encoded proteins involved in RNA processing and control of stability are imported into the plastid, although the precise way in which they are involved in RNA processing is at present unclear. The fact that a large number of RNA binding proteins have been identified in chloroplasts suggests that most mRNA molecules are stabilised by such protein complexes before they are replaced by ribosomal proteins to facilitate translation. Foremost of these RNA binding proteins are the pentatricopeptide repeat proteins (PPR), which form a large family of nuclear genes in plants and appear to be involved in RNA binding in organelles, including the plastid. Indeed, the PPR gene family in *Arabidopsis* comprises around 450 members, of which a significant proportion are involved in plastid function, probably at the level of RNA binding. PPR proteins are largely absent from prokaryotes and algae and their abundance in higher plants appears to relate to the complexity of gene regulation and possibly RNA editing. It is possible that these large numbers of PPR proteins are involved individually in specific editing or in specific transcription events during plastid biogenesis.

Plastid ribosomes and translation

Plastid ribosomes are abundant throughout the stroma of the plastid, especially in chloroplasts undergoing rapid development in leaf mesophyll cells. They can be observed in electron micrographs of chloroplasts as small dark objects (Fig. 3.7), and estimates of ribosome population sizes in chloroplasts suggest they contain around 10^5 ribosomes per chloroplast. Although the majority of the ribosomes are located in the stroma, some ribosomes are associated with the surface of the thylakoid membrane, which can also be seen in electron micrographs (Fig. 3.7). The plastid ribosome has many characteristics similar to those of bacterial ribosomes and is of a similar size, 70S, compared to ribosomes in the cytoplasm of the eukaryotic cell, which are 80S. The plastid 70S ribosomes are composed of large (50S) and small (30S) subunits, many of the protein components of which are similar to their bacterial counterparts. In the 50S subunit, there are 33 proteins, of which 8 are plastid-encoded and the rest are nuclear-encoded. In the 30S subunit, there are 25 proteins, of which 12 are encoded on the plastid genome and the rest are nuclear-encoded. A few protein components of both subunits do not have counterparts in bacterial ribosomes and are considered plastid-specific.

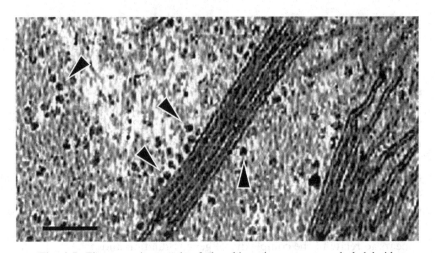

Fig. 3.7. Electron micrograph of the chloroplast stroma and thylakoid membrane showing dark stained ribosomes in the stroma (arrowed) and a line of ribosomes associated with the surface of the thylakoid membrane (arrowed). Bar = 100 nm. (http://botit.botany.wisc.edu/images/130/Plant_Cell/Electron_Micrographs/chloroplastgrana.)

These are nuclear-encoded and are termed plastid-specific ribosomal proteins (PSRP). In the 30S ribosome there are four PSRP proteins, and in the 50S ribosome there are two PSRP proteins. The prokaryotic nature of the plastid ribosomes is also shown by the fact that their translation can be inhibited by antibiotics such as spectinomycin and streptomycin, which inhibit prokaryotic-type ribosomes in bacteria, but which do not inhibit eukaryotic-type ribosomes, such as those found in the cytoplasm of the plant cell. Four ribosomal RNA molecules, which form part of the ribosomal structure, are encoded by genes on the plastid genome (Table 3.2), and which are located in the inverted repeat region (Figure 3.5), thus doubling the number of copies per genome. These genes encode three rRNA molecules that function in the 50S ribosomal subunit and one rRNA molecule that functions in the 30S ribosomal subunit.

A plastid mRNA molecule ready for translation has several features that are crucial to efficient translation; a 5′ untranslated region (5′ UTR) with a stem loop structure, a translation initiation codon, usually AUG, a translation termination codon, usually UAA, a 3′ UTR region and a stem loop structure, which confers mRNA stability. Initiation of translation involves the sequential binding of the 30S and 50S ribosomal subunits to the area around the translation start codon of the mRNA, with the aid of initiation factor proteins, imported from the cytoplasm. Elongation of the polypeptide chain then commences using tRNA molecules, the genes for which are encoded in the plastid genome and are present for all 20 amino acids (Table 3.2). Each of these tRNAs needs to be charged with the specific amino acid residues by acyl-tRNA synthetases, all of which are nuclear-encoded and are imported from the cytoplasm. The elongation process requires elongation factor proteins, which are also nuclear-encoded and those discovered so far are all homologues of prokaryotic elongation factor proteins present in bacteria. Once the ribosome complex reaches the termination codon of the mRNA, it disassociates from the mRNA molecule and the newly translated protein is released. The plastid genome uses the same genetic code as that used by nuclear genes, but although tRNAs for all 20 amino acids are plastid-encoded, several synonymous codons, that is, different codon sequences encoding the same amino acid that are used in the plastid genes, do not have the corresponding tRNA encoded in the plastid genome. Since there is no evidence for plastid tRNA import, exactly how these codons are translated is unclear, although wobble at the third base in the codon in relation to how it is read may provide a mechanism.

The observation that ribosomes can be observed associated with the thylakoid membrane suggested that the translation of plastid-encoded membrane proteins, which reside in the thylakoid membrane, could be inserted or become associated with the membrane as soon as they are translated, in a co-translational manner.

Indeed, the D1 protein, encoded by the *psbA* gene on the plastid genome, and which is a core protein of photosystem II, is translated by such membrane-bound ribosomes and assembled post-translationally into the membrane. This mechanism is facilitated by a protein, which recognises the N terminal part of the polypeptide chain as it is produced by the ribosomal complex and enables it to be inserted into the thylakoid membrane. This protein is called cpSRP54 and has a homologue in the cytoplasm, which facilitates co-translational import into the endoplasmic reticulum.

It is clear from what has been discussed in this chapter that the control of plastid gene expression and the resulting production of individual proteins encoded by the plastid genome are highly complex and contain a variety of steps, each of which could act as a point of control. These include initiation of transcription by polymerases, the extent of transcription, cistronic processing, intron splicing, RNA editing, the extent of RNA stability, translation initiation and the rate of translation. Each of these steps may be controlled by different developmental and environmental factors, providing the potential for variable control factors in different points in plastid development. A major factor in this regard is the effect of light on plastid gene expression and the translation of specific mRNA molecules, which is considered in detail in Chapter 6.

4

Photosynthesis

The process of photosynthesis as carried out by the chloroplast is arguably the most important biochemical process that occurs on the planet. Whilst many organisms gain their energy and carbon molecules by consuming other organisms in a heterotrophic manner, ultimately the basis of all food chains is the energy and carbon which is accumulated by autotrophic organisms carrying out photosynthesis. All of these organisms either contain photosynthetic chloroplasts or are single-celled organisms, such as photosynthetic bacteria, which carry out photosynthesis themselves. Such photosynthetic organisms are autotrophic in that they obtain their energy directly from the light emitted from the sun, a result of nuclear fusion. The light energy is transduced into energy stored in molecular bonds, which is then used to drive a complex series of biochemical reactions, which enables the fixation of gaseous carbon dioxide (CO_2) molecules from the air. This fixation process involves binding of the CO_2 to simple phosphorylated sugar molecules located in the stroma of the chloroplast, which eventually give rise to related phosphorylated sugar molecules which are either exported or retained in the chloroplast and give rise to biochemical pathways which synthesise sucrose or starch. In this way, plants can grow and increase in biomass by accumulating carbon molecules from the atmosphere and subsequently they synthesise a myriad of large complex carbon-based molecules. Probably the most abundant of these is the polymer cellulose, composed of long chains of glucose molecules, which forms the basic structure of plant cell walls and comprises a significant proportion of global plant biomass. The amount of photosynthetic carbon fixation carried out by photosynthetic organisms on the planet, namely higher and lower plants, algae and photosynthetic bacteria is vast, in the region of 120 gigatonnes of carbon each year. Of this, around 60% is fixed by terrestrial photosynthetic organisms and around 40% is fixed by aquatic photosynthetic organisms in the oceans.

The intimate details of the photosynthetic process, its dynamic nature and the mechanisms of its control in relation to environmental change are complex and beyond that of a short chapter on photosynthesis in a book on plastids. There are many excellent books and free articles on the web, which describe, in extensive detail, all aspects of the photosynthetic process, and some are listed in the resources at the end of this book. Thus here, we will consider a general overview of how photosynthesis works and the key elements of the process, particularly in relation to chloroplast structure.

Chlorophyll and light capture

The primary absorption of light energy by the chloroplast is carried out by the green pigment chlorophyll, the biosynthesis of which is considered in Chapter 6. Chlorophyll absorbs light energy mostly in the red and blue regions of the visible light spectrum (Fig. 4.1) and its relative inability to absorb green light confers on it its colour. In higher plants, there are two main types of chlorophyll, chlorophyll a and chlorophyll b, which only differ by the substitution of a methyl group (Chl a) for a formyl group (Chl b) in the porphyrin ring head group (see Fig. 6.6). This conversion is carried out by the enzyme chlorophyll a oxidase. Chlorophyll a and b differ slightly in their absorption spectrum (Fig. 4.1) and, in normal chloroplasts in leaves under average light conditions, they are present in a chlorophyll a/b ratio of between 2 and 4. However, significant changes in the chlorophyll a/b ratio can occur with changes in the photoenvironment of the leaf, the level of stress that it experiences and its age. Average leaves contain around 0.5 g of chlorophyll per m^2 of leaf area, which equates to about 1 picogram of chlorophyll per chloroplast in leaf mesophyll cells, equivalent to 10^8 chlorophyll molecules.

The mechanism to capture light energy and transduce its energy into molecular bonds is carried out by a series of protein complexes, which are embedded in the thylakoid membrane, with extrinsic surfaces on both the stromal and thylakoid lumen surfaces. Chlorophyll is associated with two of these thylakoid complexes, namely the two photosystems, photosystem I (PSI) and photosystem II (PSII) (Fig. 4.2), each of which contains special chlorophyll molecules at their respective reaction centres. However, most of the chlorophyll molecules present on the thylakoid

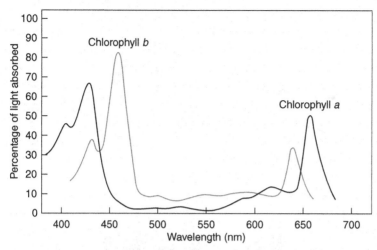

Fig. 4.1. The absorption spectra of chlorophyll a and chlorophyll b reveals how they absorb light mostly at the blue and red ends of the visible light spectrum and absorb little in the green and yellow range (500–600 nm). Consequently, chlorophyll appears as a green pigment. There are differences in the absorption spectrum of the two types of chlorophyll with chlorophyll b generally absorbing at slightly shorter wavelengths, especially toward the red end of the spectrum.

membrane are bound into arrays by association with light harvesting chlorophyll binding proteins (LHC). These arrays are termed antennae complexes and are associated with both photosystems; those attached to PSII contain chlorophyll bound to LHCII proteins and those attached to PSI contain chlorophyll bound to LHCI proteins. The PSII antennae complex associated with an individual PSII complex contains about 300 chlorophyll a and chlorophyll b molecules along with a variety of other secondary pigment molecules (Fig. 4.3). These secondary pigments are mainly carotenoid-type molecules, particularly lutein, violaxanthin and neoxanthin (see Chapter 7), whose role in the antenna complex is two-fold. Firstly, they extend the range of spectra for which light can be absorbed by the antenna complex and, secondly, they help to dissipate light energy when the light intensity is too great for the system to process all the light energy that it absorbs, and which could lead to photoinhibition of the light energy transfer and electron transfer in the reaction centres. The array of secondary pigments present in chloroplasts is revealed spectacularly in autumn when leaves senesce and chlorophyll is broken down to reveal the underlying coloured

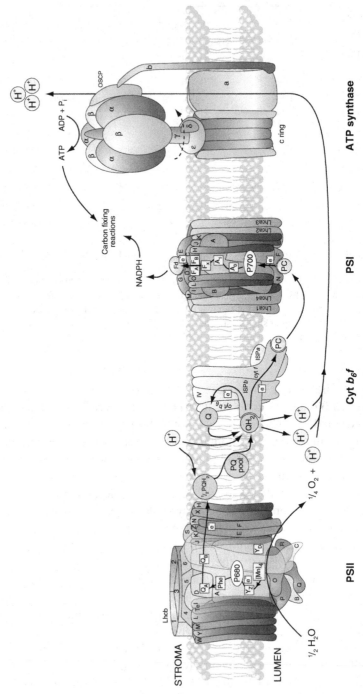

Fig. 4.2.

secondary pigments, which are mostly yellow or red and which are normally masked by the extent of the chlorophyll content of the leaf.

The actual absorption of a photon of light by a chlorophyll molecule causes an elevation in the energy level of an electron in the magnesium ion in the centre of the porphyrin ring head group. By complex photochemistry, the energy in the energised electron can be passed to a neighbouring chlorophyll molecule within the antenna complex by a process called resonance (Fig. 4.3). This results in a funnel effect of the light energy absorbed by the antennae complex in that the energy is passed between chlorophyll molecules and secondary pigment molecules until it is eventually transferred from the antennae complex to special bound chlorophyll molecules at the core of the reaction centres in both PSII and PSI. A good analogy for these antennae systems is that of a radio antenna

Caption for Fig. 4.2.

A diagrammatic representation of the major protein complexes that reside in the thylakoid membrane of the chloroplast. Four main complexes are shown: photosystem II, the cytochrome b_6f complex, photosystem I and the ATP synthase complex. Electrons are transferred between complexes in a left to right flow from photosystem II through the cytochrome b_6f complex to photosystem I. The result of electron flow is the generation of protons in the lumen as a result of water photolysis on the lumenal side of photosystem II and the plastoquinone driven movement of protons from the stroma into the lumen. Protons move out through the ATP synthase complex back into the stroma and, as they do, ATP is generated. The end point of electron transfer is the synthesis of NADPH on the stromal side of photosystem I. Most of the subunits associated with each protein complex are shown, although some are omitted for clarity. Each protein subunit is designated a letter according to the name of the gene that encodes it; for example, protein W in PSII is the product of the *psbW* gene. It must be remembered that such diagrams as this are only for understanding the basic system and that the heterogeneous distribution of the complexes in the stacked and unstacked thylakoid membranes mean that such close physical interaction of all four of these complexes rarely occurs. Labelling: LhcB1-6 – different types of light harvesting chlorophyll binding proteins, A – D1 protein, B and C in PSII are CP47 and CP43 proteins, respectively, D – D2 protein, Y_z/Y_d – tyrosine residues on D1 and D2, respectively. Phe – pheophytin, Q_A/Q_B – quinones bound to D2/D1, respectively, PQ – plastoquinone, ISP – iron–sulphur protein, PC – plastocyanin, LhcA1-4 – different types of light harvesting chlorophyll binding proteins, A/B – reaction centre core proteins of PSI, $F_X/F_A/F_B$ – iron sulphur complex, Fd – ferrodoxin, OSCP – oligomycin sensitivity conferral protein, ATP synthase chain 5. (Redrawn from figure courtesy of Jon Nield, Queen Mary University of London, http://photosynthesis. sbcs.qmul.ac.uk/nield/psIIimages/AllComplexesLarge.jpg.)

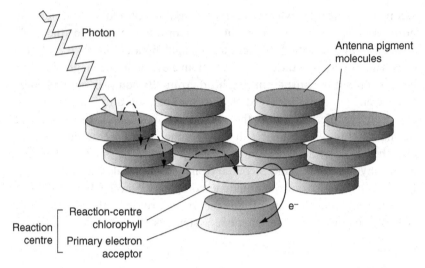

Fig. 4.3. Light-harvesting antennae complexes are associated with both photosystems and contain arrays of chlorophyll molecules associated with light-harvesting chlorophyll binding proteins (LHC). Light is absorbed by these arrays and energy moves between them by resonance transfer (dotted arrow). Eventually, the energy is transferred to the reaction centre chlorophyll, bound in the core of the reaction centre, which transfers an electron to a primary electron acceptor (solid arrow).

dish itself, which gathers radiation and focuses it to a point, the receiver (Fig. 4.3). In photosynthesis, the receiver is composed of two chlorophyll molecules, which are bound to the core proteins of the two photosystem complexes PSII and PSI (Fig. 4.2). In the absence of energy transfer from an excited chlorophyll molecule, the energy is dissipated as heat or as fluorescent light, which is red. Whilst the low level of red fluorescence emitted from chloroplasts in green leaves is normally masked by the extent of green light emitted, chlorophyll florescence can be readily seen when chlorophyll is fluoresced in a florescence or confocal microscope or when chlorophyll is extracted from leaves and illuminated.

Electron transfer and the photolysis of water

The end result of energy transfer through the chlorophyll antennae systems associated with PSII is the funnelling of excitation energy to the two special reaction centre chlorophyll molecules, termed P680, which are bound to the two core proteins of PSII, called D1 and D2 (Fig. 4.2). These D1 and D2 proteins are encoded on the chloroplast genome by *psbA* and *psbD* genes, respectively. The reaction centre is termed P680 because this is the maximal wavelength of light that the reaction centre chlorophylls can absorb. In contrast, the equivalent reaction chlorophylls in PSI are termed P700 since they can also absorb wavelengths of far red light. The reaction centre chlorophyll molecules in PSII become energised by this energy transfer, which results in the transfer of a high energy electron from a reaction centre chlorophyll to an acceptor molecule in the PSII complex called pheophytin. Pheophytin is a chlorophyll-type molecule, which lacks the central Mg^{2+} in the porphyrin head group. As a result, oxidised chlorophyll (Chl^+) is transiently produced and requires an electron source in order to become reduced and thus enable further electron transfer events to occur. The donor of electrons to Chl^+ in PSII is a tyrosine residue on the D1 protein, which itself becomes oxidised as it reduces Chl^+. In turn, the tyrosine residue is reduced by electrons derived from water molecules, which are broken down into three component parts by the process of photolysis. By using light energy and a complex cycle of photochemical reactions involving changes in the oxidation state of manganese ions, water is photolysed into hydrogen ions (protons), electrons and oxygen.

The equation is thus:

$$2H_2O \rightarrow 4H^+ + 4e^- + O_2.$$

The photolysis of water is carried out by a collection of proteins, associated with PSII, which reside on the lumenal side of the thylakoid membrane (Fig. 4.2). The physical positioning of these proteins in the thylakoid lumen is important, since it results in the protons, which arise from the photolysis reaction, being generated in the thylakoid lumenal space. The electrons generated from the photolysis reaction reduce the oxidised tyrosine residue on the D1 protein at the core of PSII and the system is prepared for further energy absorption from the antenna complex. The complex of proteins that carry out photolysis, termed the oxygen evolving complex (OEC), contains proteins both plastid-encoded and nuclear-encoded. CP43 and CP47 are both encoded by plastid genes, whereas three associated proteins, OEC33, OEC23 and OEC17 are nuclear-encoded. Also present are two subunits of cytochrome b_{559} and chlorophyll molecules, which are associated with CP43 and CP47. At the core of this complex is a cluster of manganese ions associated with Ca^{2+}, Cl^- and oxygen in a precise array. Also associated with the OEC and PSII are several smaller proteins whose exact function is unclear. Thus the entire PSII core complex with its oxygen-evolving capacity contains about 29 proteins (Fig. 4.2).

The molecule of gaseous oxygen formed by photolysis in PSII diffuses from the thylakoid membrane, out through the plastid envelope into the cytosol and through the plasma membrane and cell wall into the intracellular airspaces within the leaf from where it leaves the leaf via the stomata. Although essentially a waste product of the photosynthetic process, this oxygen provides the basis for all life on the planet, which carries out oxygen-based respiration, including the human population.

The reduction of pheophytin by electrons received from the reaction centre of PSII then sets in train a series of electron transfers, which eventually results in an electron being donated to the reaction centre of PSI. Firstly, pheophytin transfers its electrons to two quinone molecules, Q_a and Q_b, which bind to quinone binding sites on both the D1 and D2 proteins, respectively. Between the two photosystems, the electron carriers are plastoquinone, the cytochrome b_6f complex and plastocyanin (Fig. 4.2), so electrons then pass from Q_a and Q_b to reduce plastoquinone. The physical separation of the electron carriers between the two photosystems is such that the initial reduction of plastoquinone occurs on the stromal side of the thylakoid membrane whereas plastocyanin resides on the lumenal side of the thylakoid membrane. Plastoquinone is an important player in the electron transport pathway because it is the extent of reduction or oxidation of the pool of plastoquinone molecules in the membrane, which acts as a source of redox signalling which initiates a signal transduction pathway controlling both nuclear and plastid gene expression (see Chapter 6). Also associated with the transfer of electrons to and from plastoquinone is plastoquinone's ability to move protons across the thylakoid membrane from the stroma into the thylakoid lumen. Thus electron transfer through plastoquinone causes a pumping of protons into the thylakoid lumen. Together with the generation of protons from the OEC on the lumenal side of the thylakoid membrane, a significant proton concentration gradient is formed across the thylakoid membrane in the light, in the order of three pH units. The formation of this gradient is fundamental to the production of ATP by the thylakoid membrane as will be described shortly.

Plastoquinone transfers its electrons to the cytochrome b_6f complex, which is composed of cytochrome proteins f and b_6 and Rieske iron-sulphur proteins in a ratio of $2:1:1$ (Fig. 4.2). The complex also contains several polypeptides encoded by plastid genes; four polypeptides encoded by *petB*, three encoded by *petD* and one each encoded by *petA*, *petC*, *petG*, *petX* and *petL*. In turn, electrons are then transferred from the cytochrome b_6f complex to plastocyanin, a small copper-containing protein, which is nuclear-encoded, which resides on the lumenal surface of the thylakoid membrane and binds into a docking site on the cyt b_6f complex where electron transfer can occur (Fig. 4.2).

Plastocyanin then transfers electrons to PSI, which absorbs light energy using antenna complexes composed of LHCI proteins and chlorophylls in a similar way to that which occurs in PSII. PSI contains 13 polypeptides of which five are plastid-encoded and the rest are nuclear-encoded. The core of the PSI reaction centre is composed of two proteins, PSI-A and PSI-B, which are encoded by the plastid genes *psaA* and *psaB*, respectively. In a similar way to PSII, two special chlorophylls are bound at the reaction centre, called P700, which become energised by energy flow from the PSI antenna complex. Oxidised P700 then becomes reduced by the transfer of electrons from plastocyanin (Fig. 4.2). The primary acceptor of energised electrons in the core of PSI is an individual chlorophyll a molecule A_0, which passes electrons on to a secondary acceptor A_1, which are two vitamin K_1 molecules in the PSI complex, These, in turn, pass electrons to the iron-sulphur centre 4Fe–4S, which then reduces ferredoxin, a molecule of major importance in the electron transport pathway. Ferredoxin is a small nuclear-encoded protein of 10 kD containing an Fe–S group and is capable of reducing various different components on the thylakoid membrane, although the normal route is to reduce $NADP^+$ to generate the high energy molecule NADPH.

The equation is

$$2\,\text{Ferredoxin}^- + NADP^+ + H^+ \rightarrow NADPH + \text{Ferredoxin}.$$

This reaction is carried out by an enzyme, Ferredoxin $NADP^+$ oxido-reductase that is located on the stromal surface of the thylakoid membrane. NADPH is one of the two high energy final products of the light-driven electron transport part of photosynthesis and is used in various metabolic reactions in the chloroplast, in particular in the carbon fixation cycle (see later). In addition to transferring electrons to $NADP^+$, ferredoxin can also transfer electrons back to plastoquinone thereby generating a cyclic flow of electron transport, or can reduce nitrate or sulphate ions as part of the plastid's metabolism (see Chapter 7).

Ferredoxin can also reduce a family of small water-soluble proteins called thioredoxins. This reduction is carried out by the ferredoxin-thioredoxin reductase enzyme, which is a nuclear-encoded heterodimer of 30 kD, located on the stromal membrane surface. Thioredoxins become reduced at specific cysteine residues, which break a disulphydryl bridge and thus activate the thioredoxin molecule. This activation can only occur in the light whilst light-driven electron transport occurs and whilst there is a source of reduced ferredoxin. It is by using reduced thioredoxin that a variety of proteins and enzymes within the chloroplast are activated by light, since thioredoxin has the ability to reduce disulphydryl bonds in specific enzymes, thereby activating the enzyme, by virtue of changes in protein folding. Chloroplasts contain several different forms of thioredoxin, which activate different enzymes in the chloroplast, most notably enzymes involved in the Calvin cycle carbon dioxide fixation (see later).

Storing light energy as ATP

Adenosine triphosphate (ATP) is a ubiquitous molecule in cells and is used extensively as a high-energy molecule used to drive metabolic reactions. In the chloroplast, it is synthesised on the thylakoid membrane via a complex called the ATP synthase complex. A concentration gradient of protons across the thylakoid membrane forms as a result of photolysis within the thylakoid lumen and as a result of proton pumping into the thylakoid lumen via plastoquinone. Thus the concentration of protons in the thylakoid lumen is high and in the stroma it is low. The resulting pH gradient (ΔpH) across the membrane is approximately three pH units in normal light conditions, equivalent to a 1000-fold difference in proton concentration. In addition, the difference in proton concentration, together with electron transport through the membrane, generates an electrical potential across the thylakoid membrane ($\Delta\psi$), which is normally in the region of 10–50 mV. Together, these two factors generate a proton motive force, which drives the protons across the membrane from the thylakoid lumen to the stroma, in order to maintain electrical neutrality. Protons are unable to move through the membrane itself, but flow into the stroma through the ATP synthase complex, which is composed of two distinct components, each composed of different subunit polypeptides (Fig. 4.2). The CF_0 component resides in the thylakoid membrane and is made up of three subunits, a, b and c, all of which are plastid-encoded. On the stromal surface is an extrinsic CF_1 complex, which is composed of five subunits, α, β, γ, δ and ϵ, which are all plastid-encoded except for γ and δ, which are nuclear-encoded. CF_0 forms a barrel containing 12 c subunits, which defines a pore in the thylakoid membrane through which protons stream out into the stroma. As they move through the CF_0 complex, it rotates in the membrane and enables ADP and phosphate molecules to enter the active site on the CF_1 extrinsic complex. One complete rotation of the complex requires the passage of 12 protons and generates three ATP molecules, which then diffuse into the stroma. Rotation of the protein mechanism is fast, up to 1000 rotations per second, and thus ATP synthesis by the synthase complex is rapid. This mechanism is essentially the same as that used to generate ATP in the mitochondrion during respiration, and uses homologous proteins and complexes.

The ATP synthase can only make ATP if activated itself and this comes about by two main mechanisms. Firstly, the ΔpH is essential for activation and arises by exchange of water molecules within the complex. Secondly, the γ subunit is activated by reduction of a specific disulphydryl bridge using the light activated system of thioredoxin. Thus the ATP synthase only functions effectively during illumination and ATP synthesis ceases with the onset of darkness. Indeed, in darkness, the ATP synthase has activity to hydrolyse ATP and drive protons back into the thylakoid lumen, a function that ensures that the ionic environment of the thylakoid lumen is maintained and also that the system remains poised for rapid ATP synthesis with the onset of illumination.

An important aspect to appreciate of the light-induced electron transport pathway and ATP synthesis is its speed. The process of light absorption and energy transfer to reaction centres is extremely rapid, on a picosecond (10^{-12} s) timescale or faster, whereas the movement of electrons takes place in milliseconds. The movement of an electron from water in the OEC at PSII to $NADP^+$, on the extrinsic surface of PSI takes place in about 20 milliseconds (10^{-3} s).

Protein complex distribution in the thylakoid membrane

Although cartoons of thylakoid protein complexes normally show the four complexes in a sequence, as in Fig. 4.2, in reality, the four thylakoid

membrane complexes are distributed non-uniformly in the thylakoid
membrane, particularly in relation to the granal stacking. The appressed
thylakoid membranes within the granal stacks contain the majority of the
PSII complexes and their associated antenna complexes and they also
contain most of the cyt b_6f complexes. In contrast, the majority of PSI
complexes, together with their antenna complexes, are present on the
outer facing membranes at the edges of the granal stacks and in the
stromal lamellae as are the ATP synthase complexes (Fig. 4.4). Such a
structured distribution of thylakoid protein complexes is also dynamic in
that the proportioning of photosystem distribution between stacked and
unstacked membranes can alter according to environmental factors, especially in photoenvironments with high or low light levels and in relation to
short- or long-term acclimation. A further level of complexity arises from
the fact that the core of PSII operates at high oxidising potentials and is
easily damaged, especially under conditions of high light. As a result,
damaged PSII reaction centres need to be identified and replaced in the
thylakoid membrane, otherwise the efficiency of photosynthetic electron
transport will decline. The normal functional PSII reaction centre exists
as a dimer within the membrane consisting of two D1/D2 core complexes
surrounded by CP43 and CP47 proteins, antenna complexes, as well as
other associated proteins and the OEC complex. The extent of repairs to
these PSII dimers is substantial such that, at any given time, a significant
proportion of PSII complexes are undergoing repair. Consequently, in
peripheral parts of the granal stacks and in the stromal lamellae, PSII
complexes are present in a variety of heterogeneous states, including
monomer PSII complexes, monomer PSII complexes lacking CP43 and
basic reaction centre complexes containing only D1, D2 and cyt_{b559}
proteins. In addition, the size of antenna complexes varies according to
positioning within the granal stacks, with the largest antenna complexes
in the granal centre and smaller antenna complexes towards the grana
periphery.

   The major site of damage in the PSII complex, which requires replacement, is the D1 protein at the core of PSII. Its rate of turnover is very high
and is the most rapidly turned over protein on the thylakoid membrane.
The damage-induced repair cycle of PSII and replacement of damaged D1
proteins is complex and involves phosphorylation and dephosphorylation
of components of the PSII complex including D1 itself, monomerisation
of the PSII core and subsequent migration from the granal stacks out into
the stromal lamellae. Once in the stromal lamellae, PSII is partly disas-
sembled and the damaged proteins are degraded, to be replaced by new

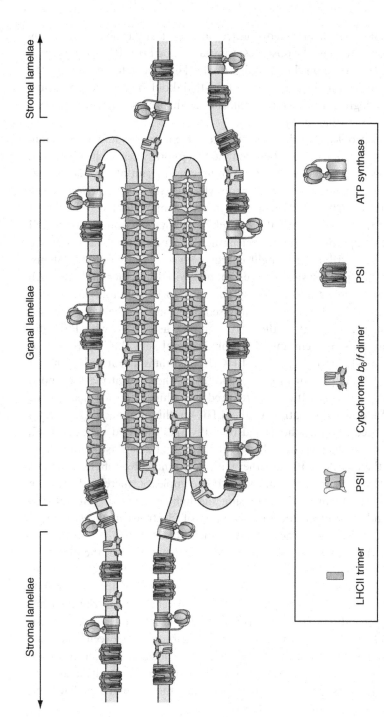

Fig. 4.4. The distribution of the thylakoid membrane complexes in the thylakoid membrane is heterogeneous in that complexes such as PSII are much more abundant in the granal stacks, whereas complexes such as PSI and ATP synthase are much more abundant in the stromal lamella and in the unappressed surfaces of the grana. (Redrawn from Allen JF, Forsberg J. Molecular recognition in thylakoid structure and function. *Trends in Plant Science* 6, 317–326. © Elsevier 2001.)

Stromal lamellae

Granal lamellae

Stromal lamellae

LHCII trimer   PSII   Cytochrome b_6/f dimer   PSI   ATP synthase

D1 proteins, which are incorporated in a co-translational manner by interaction with plastid ribosomes, translating the D1 mRNA transcribed from the *psbA* gene on the plastid genome. Thus the heterogeneity of the structure of PSII in different parts of the thylakoid membrane represents, to a large degree, different stages in the repair process of damaged PSII complexes (see Chapter 8).

Another problem faced by the thylakoid membrane is how to maintain a constant flow of electrons between the two photosystems when PSII and PSI absorb slightly different spectra of light as a result of the characteristics of their reaction centre chlorophylls. Because the reaction centre chlorophylls P680 in PSII can only absorb a maximum wavelength of red light, whereas the special chlorophylls at the reaction centre of PSI can absorb up to 700 nm wavelength, there needs to be a system to balance the distribution of light energy between the two photosystems, since the two photosystems are essentially wired together in series such that electron flow through the system must occur without logjams or pileups. In order to facilitate optimised electron flow between the two photosystems, one part of the LHC complex associated with PSII and containing LHC protein and chlorophyll is mobile within the thylakoid membrane. This mobile pool of LHCII constitutes about 15%–20% of the total LHCII present in the thylakoids. If the pool of plastoquinone molecules becomes sufficiently reduced and thus electrons need to be passed on down the pathway, the LHC in this mobile PSII antenna complex becomes phosphorylated by a thylakoid located kinase, STN7, the activity of which is controlled by the redox state of plastoquinone. As a result, the mobile LHC, together with its bound chlorophylls, migrates from PSII and becomes attached to PSI complexes, thereby decreasing the light capture of PSII, increasing the light capture of PSI antenna system and promoting the through flow of electrons in the system. If the photoenvironment then changes such that there is a shortage of electrons in the plastoquinone pool, and such that most of the plastoquininone is oxidised, then the LHCII kinase is switched off, LHCII becomes dephosphorylated by a phosphoprotein phosphatase and the mobile LHCII antenna returns to photosystem II to rebalance energy distribution. Movement of LHCII in this way is termed a state II–state I transition, and acts to fine-tune energy distribution between the two photosystems. Since the two photosystems are located differently within the membrane, the initial movement of the mobile LHC is outwards from the granal stacks toward the periphery and the stromal lamellae, and when dephosphoryated its path is reversed.

Fixation of carbon dioxide and the Calvin cycle

The energy stored in the pool of ATP and NADPH molecules synthesised by the light-driven electron transport on the thylakoid membrane is used in a variety of enzymatic reactions in the stroma of the chloroplasts, but its major use is to drive a series of biochemical reactions, which constitute a cycle by which carbon dioxide is fixed and resultant sugar molecules are stored as starch or exported from the chloroplast into the cell. The cycle is normally referred to as the Calvin cycle, named after Melvin Calvin who first described the system but is also known as the reductive pentose phosphate pathway or the photosynthetic carbon reduction cycle. This cycle is the fundamental process by which CO_2 is assimilated in all photosynthetic organisms, including photosynthetic prokaryotes, and which is normally referred to as C_3 photosynthesis, as the first major product of CO_2 fixation is a molecule containing three carbon atoms. In many texts, these parts of photosynthetic biochemistry have continually been referred to as the dark reactions or the light-independent phase of photosynthesis, in contrast to the light-dependent reactions of thylakoid electron transport. This is blatantly wrong since, although the biochemistry of the Calvin cycle does not need light directly in order to function, many aspects of its biochemistry are activated by light and it does not proceed at all in the dark. Thus all aspects of the photosynthetic process are basically dependent upon light.

The Calvin cycle is complex and contains 11 enzymes, which function in 13 different steps to enable the cycle to function (Fig. 4.5). The primary step of fixing carbon dioxide is carried out by the enzyme ribulose 1,5 bisphosphate carboxylase/oxygenase (RUBISCO), in which CO_2 combines with a phosphorylated sugar, ribulose 1,5 bisphosphate (RuBP) to form 3-phosphoglyceric acid (3-PGA). 3-PGA contains three carbon atoms and hence RUBISCO initiates C_3 photosynthesis. The Calvin cycle then functions to export a proportion of the fixed carbon for metabolic synthesis and to regenerate more RuBP substrate so that the fixation reaction can occur subsequently, in a cyclical manner. RUBISCO is a famous enzyme in that its genetics have been very well studied, as has its biochemistry and structure. The RUBISCO holoenzyme is composed of 16 subunits, totalling a molecular mass of around 500 kD. Eight are large subunits (LSU), encoded by a single gene on the plastid genome and eight are small subunits (SSU), encoded by a family of nuclear genes, and imported from the cytosol. The pattern of expression of small subunit genes is complex and varies between different types of tissue within a

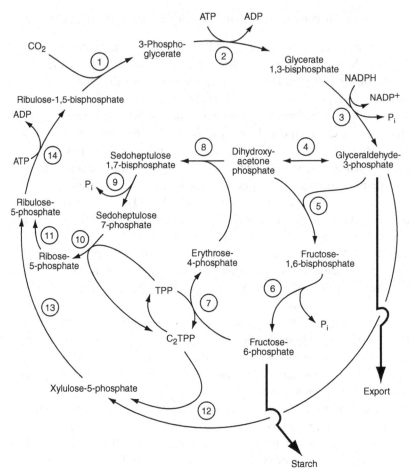

Fig. 4.5. A schematic diagram of the Calvin cycle, which functions in the chloroplast, showing how CO_2 is fixed by the enzyme RUBISCO and products formed subsequently in the form of glyceraldehyde 3-phosphate are exported from the plastid to make sucrose. In addition, fructose 6-phosphate can be used to synthesise starch. The rest of the complex cycle involving many enzymatic reactions functions to resynthesise ribulose-1,5-bisphosphate so that the cycle can continue. Enzymes are: (1) ribulose-1,5-bisphosphate carboxylase/oxygenase (RUBISCO), (2) phosphoglycerate kinase, (3) triose phosphate dehydrogenase (NADP glyceraldehyde phosphate dehydrogenase), (4) triose phosphate isomerase, (5,8) Aldolase, (6) fructose bisphosphatase (7,10,12) transketolase, (9) sedoheptulose bisphosphatase, (11) ribose-5-phosphate isomerase, (13) ribulose-5-phosphate-3-epimerase, (14) ribulose-5-phosphate kinase. TPP = thiamine pyrophosphate. (Redrawn from Lawlor D. (2001 *Photosynthesis*. Fig. 7.1 Third Edition). Garland publishing.)

plant, and varies proportionately between different family members under differing environmental conditions, especially light. Thus the RUBISCO holoenzyme can be built of different combinations of SSU polypeptides from different gene family members and this may represent an adaptation of the RUBISCO enzyme to environmental change.

SSU polypeptides are imported into the plastid after translation on cytoplasmic ribosomes, whereas LSU polypeptides are synthesised on plastid ribosomes and subsequently interact with chaperone proteins, BSD2 and cpn60, to facilitate their folding and enable efficient construction of the RUBISCO holoenzyme.

Since the RUBISCO holoenzyme is composed of subunits from two different genetic compartments, there must be some degree of coordination between the production of the two types of subunit in order to enable efficient construction of the holoenzyme in the plastid. Feedback control onto the production of the two different subunits as a monitor of the availability of its partner is different for the two subunits. For LSU, free LSU subunits, which are not bound into RUBISCO complexes, have a negative effect on the continued translation of LSU subunits by chloroplast ribosomes. Thus the presence of unassembled LSU subunits reduces further translation of LSU mRNA until a supply of SSU is imported into the chloroplast. If LSU subunits are in short supply, then excess SSU polypeptides appear to be degraded within the chloroplast if they remain unassembled.

The active site for the CO_2 fixation reaction by RUBISCO is on a lysine residue on the large subunit, so eight carboxylation reactions can occur simultaneously on a single RUBISCO holoenzyme. The role of the SSU in the reaction is largely regulatory, but SSU also maintains the architecture of the large subunits in the holoenzyme complex. The active site for CO_2 fixation on each of the RUBISCO large subunits in the holoenzyme needs to be activated before the carboxylation reaction can proceed. The activation process is complex, requiring binding of Mg^{2+} and a CO_2 molecule, which is a different CO_2 from that used for carboxylation. In addition, various intermediates of the carboxylation reaction bind to the active site in an inhibitory manner, including the substrate RuBP, thereby preventing activation from occurring. Such inhibitors are removed by the enzyme RUBISCO activase, which is nuclear-encoded and was first discovered in an *Arabidopsis* mutant, which could not activate its RUBISCO enzyme. RUBISCO activase itself is an abundant protein constituting 2%–5% of total leaf protein, but its activity may be a control point for carbon fixation, especially under conditions of low light at the beginning of the day.

RUBISCO is a remarkably poor enzyme in fixing CO_2, even though it represents presumably the best that evolution could conjure for this reaction. The low K_m for the CO_2 fixation reaction means that high concentrations of the enzyme in the stroma are required to maintain a reasonable rate of carboxylation. Thus the concentration of the enzyme in the stroma is high, at around 4 mM, which is a remarkably high concentration for a biological enzyme. The concentration of RUBISCO is such that it makes the stroma compartment very viscous and in some situations can form crystalline arrays within the stroma, visible with the electron microscope. The concentration of the RUBISCO protein is such that it can comprise up to 50% of total leaf protein and is considered to be leaf storage protein, because of its abundance. Rough calculations suggest that there is approximately 10 kg of the enzyme for each person on the planet, making it probably the most abundant protein in the world!

Once activated, RUBISCO can carry out the carboxylation reaction, which joins CO_2 to RuBP in a complex reaction involving a six-carbon intermediate molecule, which rapidly decays to form two three-carbon molecules of PGA. Once CO_2 is fixed, a subsequent set of enzymatic reactions take place to ensure that sufficient RuBP can be resynthesised to enable the fixation reaction to continue and also that some of fixed carbon can leave the Calvin cycle to be stored or be used in metabolic synthesis of other molecules. The Calvin cycle requires energy and reductant to drive some of the enzymatic reactions and these are derived from the ATP and NADPH synthesised by the light-driven electron transport pathway on the thylakoid membrane. Firstly, 3-PGA is phosphorylated to produce glycerate 1,3-bisphosphate, a reaction that requires ATP (Fig. 4.5) and glycerate 1,3-bisphosphate is subsequently converted to glyceraldehyde-3-phosphate (GAP), a reaction that requires reduction by NADPH. GAP interconverts with dihydroxyacetone phosphate (DHAP), and both GAP and DHAP are referred to as triose phosphate. In considering the fixation of six carbon dioxide molecules, five-sixths of the triose phosphate molecules generated as a result are used to regenerate ribulose bisphosphate molecules so that CO_2 fixation can continue and one-sixth is available for export to the cytosol and leaves the Calvin cycle via the triose phosphate/inorganic phosphate transporter in the inner envelope membrane and is used in the cytosol to synthesise sucrose (see Chapter 7). Depending on the type of plant species and the environmental and physiological conditions of the chloroplast, starch can be synthesised within the chloroplast from fructose-6-phosphate. This starch is transitory and is normally broken down during the dark period and the products exported (see Chapter 2).

All 11 enzymes in the Calvin cycle, with the exception of RUBISCO LSU, are nuclear-encoded and imported into the chloroplast (Fig. 4.5). Their individual activities are controlled by a variety of factors, making the overall control of the Calvin cycle extremely complex. However, four of the enzymatic steps are controlled by light (Fig. 4.5), using the thioredoxin system to activate the enzymes as described earlier. Thus the Calvin cycle will only fix carbon dioxide, export triose phosphate and regenerate RuBP in the light and will quickly stop in darkness, not least because the supply of ATP and NADPH declines rapidly. It is important to realise that, in the light, the ATP and NADPH synthesised by the light-driven electron transport system is utilised by the Calvin cycle almost immediately. The final step of the cycle in regenerating RuBP also utilises ATP, to convert ribulose-5-phosphate into RuBP, which is then ready to accept another CO_2 molecule by virtue of RUBISCO carboxylation. Characterisation of these various Calvin cycle enzymes, other than RUBISCO, has proved difficult since it was problematic to isolate them individually and it became clear that these enzymes are not present as individual molecules that diffuse through the stroma but are present as complexes which are associated with the surface of the thylakoid membrane and contain several enzymes, intimately arranged together. Exactly how these enzymes are complexed together is unclear, although a complex containing the enzymes ribose-5-phosphate isomerase and glyceraldehyde-3-phosphate dehydrogenase, together with a small non-enzymatic protein called CP12 has been shown to be associated with thylakoid membranes.

An apparent evolutionary consequence of the design of the RUBISCO enzyme and its ability to carry out the carboxylation reaction is that oxygen can also bind at the active site on the large subunit and it competes with CO_2 molecules. Thus RUBISCO is also an oxygenase enzyme, which results in RuBP combining with oxygen to form phosphoglycolate. This is a highly wasteful reaction that not only uses RuBP molecules, which could have been carboxylated but also occupies active sites preventing carboxylation. As a result of the production of phosphoglycolate, a complex set of biochemical reactions called photorespiration has evolved to retrieve the carbon lost by phosphoglycolate formation. This series of reactions involves interplay between the chloroplast, peroxisome and the mitochondrian and utilises energy. A major problem for the chloroplast is that, at low temperatures, the competition by O_2 for the CO_2 active site is moderate and the chloroplast can cope with using photorespiration to compensate. However, with increasing temperature, the availability of O_2 compared with CO_2 at the active site increases due to solubility differences between the gases as well as a more active oxygenase activity at higher temperatures. Consequently, plants growing at high temperatures would be severely compromised by RUBISCO oxygenase activity. Many plants growing at high temperatures have developed a mechanism for increasing the concentration of CO_2 at the active site in RUBISCO in a system termed C_4 photosynthesis, the chloroplast-related aspects of which are considered in Chapter 8.

5

Plastid import

During the development of the endosymbiotic relationship between the invading 'plastid' and the recipient prokaryotic cell through the course of evolution, many changes occurred to the invading 'plastid'. As we have seen, one major change has been the extensive movement of genetic information from the plastid's own genome into the nuclear genome. As a result of these events, the cell is faced with a major problem; namely, how those proteins now encoded by the cell's nucleus get back into the correct place within the plastid in order to allow the plastid to function effectively. Since around 95% of the proteins that are present in the mature chloroplast in the cells of present-day plants are encoded by genes in the nucleus, this problem becomes significantly more than just routing the occasional protein and it constitutes a major flow of protein trafficking within the cell. Plastids import much more than proteins. In all cells, plastids play a major role in metabolic biochemistry and synthesise many important molecules, which are utilised in other parts of the cell (see Chapter 7). Thus a wide variety of molecules other than proteins are imported and exported by the plastid in its normal course of biochemical function. Foremost amongst these are the end products of photosynthesis as well as lipids, amino acids and various other intermediates in biochemical pathways.

A complication which has to be overcome in trafficking molecules into and out of the plastid is that the boundary of the plastid is a double membrane, composed of the outer envelope membrane and the inner envelope membrane, with a distinct compartment in between; the envelope lumen. Thus all molecules have to traverse this boundary effectively in order to enter or leave the plastid. Once within the plastid, there are several different compartments into which the imported molecules can reside and function. They can either remain in the stroma, be imported across the thylakoid membrane into the thylakoid lumen or even be incorporated into the thylakoid membrane itself. In addition, many proteins are incorporated into the plastid envelope itself or reside in the envelope lumen. Consequently, evolution required the development of efficient mechanisms to be put in place to ensure that the proteins, now encoded in the nucleus, were targeted back to the correct compartment in the plastid, so that the plastid can function effectively.

Plastid transit peptides

In order for a protein to be able to enter the plastid completely, it must contain some information within its amino acid sequence, which allows import specifically into the plastid stroma. In the vast majority of proteins, which end up in the stroma or within the thylakoid lumen, this information is encoded in a small stretch of amino acids at the N-terminus of the protein, called the plastid transit peptide. Plastid transit peptide sequences were first discovered when the sequences of mature functional proteins in the plastid were compared to their sequences when translated on cytoplasmic ribosomes. It was found that, in each case, the mature protein is slightly shorter and is missing the transit peptide sequence, which in most cases is cleaved off from the protein after it enters the plastid. Consequently, the mature protein is shorter than the precursor protein, which was translated by the cytosolic ribosomes. Expectations that the amino acid sequence of plastid transit peptides would be conserved between precursor proteins were quickly dashed once transit peptide sequences from different proteins were compared. Plastid transit peptides show little amino acid sequence homology to each other, but display an array of distinct biochemical features of the amino acids within them, which appear more important than the actual amino acid sequence itself. Thus, transit peptides vary significantly in their sequence of amino acids (Fig. 5.1) and are very diverse among different plastid targeted proteins, varying in length between 20 and 150 amino acids. However, one major biochemical feature of the transit peptide sequence is its overall positive charge at cytoplasmic pH as well as having an enrichment in the hydroxyl amino acids serine and threonine. There is evidence that there are three functional domains within a transit peptide sequence consisting of ten amino acid residues at the N-terminal, which are uncharged and commencing with MA and ending with G/P, a central domain with no acidic residues but enriched in serine

(1) MASISSSVATVSRTAPAQANMVAPFTGLKSNAAFPTTKKANDFSTLPSNGGRVQC*
(2) MASSTMALSSTAFAGKAVNVPSSSFGEARVT*
(3) MATITGSSMPTRTACFNYQGRSAESKLNLPQIHFNNNQAFPVLGLRSLNKLHVRTARAT
SGSSDTSEKSLGKIVC*

Fig. 5.1. A comparison of the amino acid sequence of plastid transit peptide sequences at the N-terminus of proteins that are imported into the plastid. (1) RUBISCO small subunit from sunflower (*Helianthus annus*); (2) light-harvesting chlorophyll binding protein (LHC) from maize (*Zea mays*); (3) granule-bound starch synthase 1 from pea (*Pisum sativum*). The asterisk at the end of each sequence shows where it is cleaved off from the rest of the precursor protein to give a mature protein lacking the transit peptide sequence.

and threonine, and a C-domain enriched in arginine. These biochemical features appear sufficient to enable the protein to be recognised by the plastid envelope membrane receptors.

The strategy of using amino acid sequence information to target proteins to a particular cellular organelle is not confined to the plastid, since different types of sequence target proteins to many other cellular compartments such as the mitochondrion, the nucleus and the endoplasmic reticulum. A crucial aspect of the plastid transit peptide sequence and the machinery that recognises it at the boundary of the plastid is that it works correctly and that no mistargeting to other organelles in the cell occurs. During normal cell function, it would appear that mistargeting does not occur and that all proteins are faithfully targeted to their correct compartments. Even though targeting is carried out correctly, there are some proteins which carry a transit peptide sequence that enables their import into both the plastid and the mitochondria compartments, by a process called dual targeting. Thus in such a case, the pool of molecules synthesised by cytosolic ribosomes is split and some molecules end up inside the mitochondria and the rest end up inside the plastids within the cell. In addition, there are a few proteins which enter the plastid without a transit peptide sequence, suggesting that an alternative plastid import pathway may yet be discovered. However, overall the vast majority of plastid imported proteins utilise a transit peptide sequence, which interacts with import complexes on the chloroplast envelope membranes.

How does plastid import work?

Let us consider the pathway of a protein which is destined to be imported into the stroma of the chloroplast. When made on cytosolic ribosomes, the precursor protein contains a transit peptide sequence at its N-terminus. As it emerges from the ribosome, it interacts with cytosolic components, which aid the import process. These cytosolic components of the import process are called chaperones. Chaperone proteins interact with precursor proteins destined for plastid import and prevent them folding prematurely, prior to their interaction with the import complexes on the chloroplast envelope membrane. The import complexes require that proteins to be imported are in a linear, unfolded state and the chaperone proteins facilitate this. The class of chaperones involved in this process are related to 70 kD heat shock proteins (Hsp70), and the transit peptide sequences of the majority of proteins to be imported by the plastid contain a binding site for DnaK, the binding site for the Hsp70 homologue in *E. coli*. After interacting with its chaperone, serine and threonine residues within the transit peptide are phosphorylated by a cytosolic serine/threonine kinase enzyme and the protein and its chaperone interact with a guidance complex, composed of a class of proteins called 14-3-3 proteins, which recognise the phosphorylated serine/threonine residues.

The plastid envelope membranes contain two distinct protein complexes which control the import of proteins into the plastid. One complex is associated with the outer plastid envelope membrane and is called the Toc complex and the other is associated with the inner plastid envelope membrane and is called the Tic complex (Fig. 5.2). Both of these complexes are termed translocons, which refers to a complete set of proteins that enables translocation of a precursor protein destined for the plastid stroma or other internal plastid compartment into the plastid. A component of the Toc complex, Toc64, probably interacts with the chaperone and the guidance complex to aid docking of the precursor protein during the first phase of the import process.

The Toc translocon complex of the outer envelope membrane is made up of three core proteins, Toc159, Toc75 and Toc34, which are so-called because of their relative molecular masses. Together, they form a complex, which recognises the precursor protein destined for the plastid, and translocates it across the outer envelope membrane. The core Toc complex forms a ring structure with a width of about 130 Å. The entire complex has a mass of about 500 kD and contains a single Toc159 protein, four Toc75 proteins and four or five Toc34 proteins. It is the Toc75 protein that actually forms the pore of the complex, whereas its associated proteins, Toc159 and Toc34, play a role in reception of the precursor protein and interaction with the transit peptide sequence. Both of these proteins contain a GTP binding site and both can be

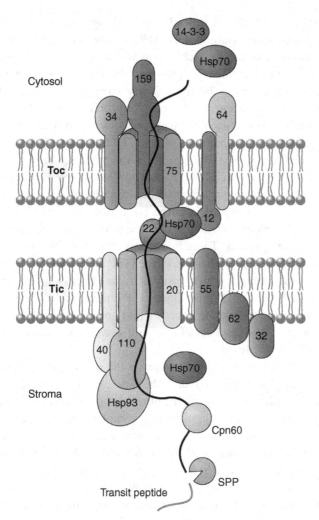

Fig. 5.2. Diagrammatic representation of how a plastid-targeted protein synthesised on cytosolic ribosomes gets imported across the two plastid envelope membranes by the Tic Toc import complexes. The protein is imported in a linear form and threads through the Toc complex in the outer envelope membrane and then through the Tic complex in the inner envelope membrane. In addition to the Tic and Toc proteins shown, which are numbered according to their molecular weights in kD, several chaperone proteins assist in the process, particularly in maintaining the imported protein in a linear conformation. These are Hsp70, Hsp93 and Cpn60. As the transit peptide sequence enters the stroma, it is cleaved by the stromal processing peptidase (SPP), to give rise to the mature protein. (Redrawn from Jarvis P, Robinson C. Mechanisms of protein import and routing in chloroplasts. *Current Biology* 14, R1064–R1077. © Elsevier 2004.)

phosphorylated, which is a crucial step in recognition of the precursor protein. This phosphorylation of Toc159 and Toc34 is carried out by two different kinase enzymes, which reside within the outer envelope but Toc159 and Toc34 cannot be phosphorylated when they have GTP bound. Toc34 appears to be the primary receptor protein of the Toc translocon and recognition of the precursor protein by Toc34 is enhanced by the precursor protein itself being phosphorylated on the transit peptide by the cytosolic serine/threonine kinase. GTP binding to Toc34 is crucial for recognition and subsequent binding of the precursor protein to Toc34 stimulates GTPase activity. Thus, phosphorylation and GTP binding to Toc34 acts as a molecular switch, which can activate or deactivate the protein. The third major protein of the Toc translocon is Toc159, which is also a GTP-binding protein and interacts with the precursor protein early on in the import process. There is evidence that Toc159 acts as a GTP-driven motor, such that hydrolysis of the GTP bound to Toc159 causes a conformational change, which results in the precursor protein being pushed through the Toc75 pore. Thus, GTP binding and hydrolysis make the Toc translocon a dynamic structure and one in which the association of Toc34 with the Toc75/Toc159 core unit is weakened when the Toc34 bound GTP is hydrolysed.

The role of the intermembrane space in the import process is unclear and it is generally considered that the Toc complex interacts with its equivalent complex in the inner membrane, the Tic complex, such that imported proteins pass through the two complexes together and do not properly enter the lumen between the outer and inner membranes. Sites where the outer and inner envelope membrane are held closely together are called contact sites and it is presumed that interaction between Toc and Tic complexes at these points enable tight membrane interaction. Toc12 and Tic22 proteins may facilitate the interaction between the two complexes across the envelope lumen (Fig. 5.2).

The structure of the Tic translocon on the inner envelope membrane is less clearly established, compared with the Toc complex, but it contains at least seven proteins; Tic110, Tic62, Tic55, Tic40, Tic32, Tic22 and Tic20. Tic110 is the largest and in association with Tic20 probably forms the core channel on the inner envelope membrane through which the imported protein travels, spanning the inner membrane and having domains in both the envelope lumen and in the stroma (Fig. 5.2). The part of the Tic110 protein that resides in the stroma can recruit chaperone proteins from the stroma, such as Hsp93, to bind to newly imported proteins as they are imported. Tic62 sits in the inner envelope membrane, has two N-terminal transmembrane helices and, at the C-terminus, which extends into the stromal compartment, it has a binding site for ferrdoxin-NAD(P)$^+$ oxidoreductase (FNR). FNR catalyses the last step in the photosynthetic electron transport chain on the thylakoid membrane and its redox state is a measure of the redox state of the chloroplast (see Chapter 4). This suggests that the activity of the Tic import complex can be modulated by the redox state of the chloroplast and this factor could control the import of those nuclear-encoded proteins, whose expression in the nucleus is also controlled by the redox state of the chloroplast (see Chapter 6). Of the remaining proteins in the Tic complex, Tic55 and Tic32 appear to play a role in regulating the translocation process, particularly in relation to redox sensing, since Tic32 possess a binding site for NAD(P)$^+$.

Consequently, the actual import process of a precursor protein and its passage through the Tic/Toc complexes occurs in three stages. Firstly, the precursor protein, associated with a chaperone, makes contact with the receptor proteins of the Toc complex. This is a reversible process, which is energy independent. Secondly, the precursor protein moves through the opening Toc complex as an unfolded protein and becomes deeply inserted in the complex. The precursor protein is now docked in an irreversible state, which requires energy in the form of ATP in the intermembrane space and GTP. The final third stage is the movement of the precursor completely into the stroma and cleavage of the transit peptide by the stromal processing peptidase (SPP). This last stage requires higher concentrations of ATP in the stroma, but in circumstances in which ATP and GTP are not limiting, the precursor protein probably moves through the import process in one continuous movement.

As the imported protein emerges into the stroma from the Tic complex, the transit peptide sequence is immediately recognised by the stromal processing peptidase enzyme (SPP), which cleaves the transit peptide from the emerging protein (Fig. 5.2). This enzyme recognises the amino acid sequence around the C-terminal area of the transit sequence and cleaves off the transit peptide sequence precisely, which is subsequently degraded by another enzyme, an ATP-dependent metalloprotease, termed the transit peptide subfragment degrading enzyme. Interestingly, the SPP enzyme also functions in the mitochondrion where it cleaves off transit peptides, which facilitate import into the mitochondrion. Since the SPP is encoded by a nuclear gene, it itself is targeted to both the plastid and mitochondrion organelles by a transit peptide which facilitates dual targeting.

Variation in import complex proteins

A novel aspect of import complex proteins was discovered when the complete gene sequence of *Arabidopsis* was analysed for those genes encoding Tic and Toc components. It was found that, in the *Arabidopsis* genome, there are two different genes encoding Toc34 proteins, called *Toc33* and *Toc34*, and four different genes encoding Toc159 proteins, which led to the idea that these genes encode different isoforms of these two proteins, which could relate to different Toc complexes importing different subsets of precursor proteins. By mutating these individual genes in *Arabidopsis* and analysing the phenotype of the plastids in the mutant plants, it became clear that different Toc159 proteins are involved in importing different types of proteins into the plastid. One group of Toc159 proteins function in complexes, which import proteins involved primarily with photosynthesis in developing chloroplasts, whereas other Toc159 proteins are involved in complexes importing proteins in non-photosynthetic plastid types such as those in roots. Mutations in the genes of these two different Toc159 groups only affect those plastid types with which they are associated for import. The evolution of such a variation in import complex heterogeneity is probably due to the fact that, during early chloroplast development, there is an avalanche of protein import of those proteins required to make the photosynthetic complexes on the thylakoid membrane, the amount of which far outweighs that of less abundant housekeeping proteins within the plastid. Consequently, in the outer envelope membrane of the young chloroplasts, there are two distinct types of Toc complex, which import different classes of precursor proteins, which presumably is a more efficient strategy for developing chloroplasts. In the Tic complex, all of the proteins are encoded by single nuclear genes, except for Tic32, Tic20 and Tic22, which have two nuclear genes each, although the relevance of this duplicity is not known.

How do envelope proteins get into the plastid envelope?

Proteins which reside in the outer and inner envelope membranes and those in the intermembrane lumen are a special class of imported proteins in that they do not need to go through the Tic/Toc complex system. This class of proteins includes, of course, the members of the Tic and Toc complexes themselves. However, several of the Tic complex components have standard transit peptide sequences, which would allow them passage through the Toc pore after which they insert into the inner membrane. Toc75 is different in that, downstream of its transit peptide sequence, it possesses a second sequence containing many glycine residues, which is thought to act as a stop-transfer domain. This results in the translocation process of Toc75 stopping when the protein is inside the complexes and the protein moves sideways into the membrane. Such a mechanism is similar to that which occurs during the import of hydrophobic membrane proteins into the endoplasmic reticulum membrane. In contrast, the Toc34 protein of the outer envelope Toc complex has no cleavable transit peptide sequence and may self-insert into the outer envelope membrane without interaction with the Tic Toc mechanism.

Targeting inside the plastid

Proteins that are imported initially into the plastid stroma can have a variety of final destinations. Many are soluble stromal proteins, which are folded correctly in the stroma with the aid of chaperone proteins and then function in the stroma or contribute to complexes on the surface of the inner envelope membrane or the surface of the thylakoid membrane. These would include enzymes of the Calvin cycle, other biosynthetic enzymes and enzymes associated with DNA and RNA metabolism, many of which will become associated with nucleoids. Alternatively, many imported proteins are targeted to the thylakoid membrane, either to become inserted into the membrane itself or imported into the lumenal space within the thylakoid. The thylakoid membrane contains several large protein complexes, which enable light capture and electron trans-port resulting in ATP synthesis in the process of photosynthesis (see Chapter 4). The lumen contains many proteins, including plastocyanin, which is involved in electron transport between the thylakoid complexes and has been well characterised in terms of its import pathway. Since it resides in the thylakoid lumen, it has to be transported across the outer and inner plastid envelope membranes as well as being transported across the thylakoid membrane itself. Proteins such as plastocyanin have a two-part transit peptide sequence which enables this complex passage (Fig. 5.3). The first part has properties of a normal stromal targeting transit peptide and is cleaved off by the SPP on entry into the stroma. This reveals a second N-terminal sequence, which enables entry into the thylakoid lumen. The translocation process across the thylakoid mem-brane is accomplished by one of two distinct mechanisms.

(a)

```
MATVASSAAVAVPSFTGLKASGSIKPTTAKIIPTTTAVPRLSVKASLKNVGAAVVATAAAGLLAGNAMA QD*
VEVLLGGGDGSLAFLPGDFSVASGEEIVFKNNAGFPHNVVFDEDEIPSGVDAAKISMSEEDLLNAPGETYKVTLTEKGTYKFYCS
PHQGAGMVGKVTVN
```

(b)

Sec-type signal peptides

```
(1)     --CADAAKMAGFALATSALLVSGATA

(2)     --LSASIKTFSAALALSSVLLSSAATSPPPAAA

(3)     --PFSAVKPFFLLCTSVALSFSLFAASPAVESASA
```

(c)

TAT signal peptides

```
(5)     --SDAAVVTSRRAALSLLAGAAAIAVKVSPAAA

(6)     --GDAVAQAGRRAVIGLVATGIVGGALSQAARA

(7)     --VQVAPAKDRRSALLGLAAVFAATAASAGSARA

(8)     --KEQSSTTMRRDLMFTAAAAAVCSLAKVAMA
```

Fig. 5.3. (a) The amino acid sequence of the thylakoid lumen protein plastocyanin with its N-terminus bipartite transit peptide sequence. The first part of the sequence is the transit peptide enabling passage through the plastid envelope membrane and the second part enables import across the thylakoid membrane by the Sec pathway, which is conferred by a conserved K residue (shown in bold) and an AXA sequence (underlined), where in this case X=methionine. The cleavage by the thylakoid processing peptidase (TPP) is shown by the asterisk, downstream of which is the amino acid sequence of the mature plastocyanin protein. The exact place in the bipartite transit peptide where the stromal processing peptidase (SPP) cleaves is not known exactly but is somewhere between the start and the conserved K residue. (b) Comparisons of the transit peptide sequences, which allow transport across the thylakoid membrane by the Sec pathway and comparable with the equivalent sequence in plastotocyanin in (a). The sequences are aligned at the conserved lysine residue (K) and the AXA sequence in each is underlined. (1) wheat 33 kD protein of the oxygen evolving complex; (2) barley photosystem I subunit F; (3) *Arabidopsis* Deg protease enzyme. (c) Comparisons of the transit peptide sequences which allow transport across the thylakoid membrane by the TAT pathway, lined up by the conserved double arginine residues (**RR**). The conserved AXA groups at the end are underlined. (4) wheat 23 kD protein of the oxygen evolving complex; (5) maize 16 kD protein of the oxygen evolving complex; (6) barley photosystem I subunit N; (7) *Arabidopsis* photosystem II subunit T.

The Sec pathway utilises proteins which are similar to those which export proteins across the plasma membranes of bacteria. SecA is an ATP-driven translocation motor and the channel in the thylakoid membrane is composed of SecY and SecE proteins. This system recognises the second part of the two-part transit peptide sequence, which is itself composed of three domains: a positively charged amino-terminal domain, a hydrophobic core domain and a polar carboxy-terminal domain. A conserved lysine residue is also important (Fig. 5.3). Proteins that use the Sec pathway are moved across the thylakoid membrane in an unfolded linear conformation. The C-terminus of the second transit peptide sequence contains an Ala–Xaa–Ala consensus sequence, which is recognised by a thylakoid processing peptidase (TPP) within the thylakoid lumen and cleaves off this second transit sequence leaving the mature functional protein (Fig. 5.3). The TPP is a serine protease and is similar in its cleavage specificity to that of bacterial signal peptidases.

A second mechanism for import into the thylakoid lumen is that of the Tat pathway, so-called because the mechanism recognises twin arginine residues in the second part of the transit peptide sequence, the first part having been cleaved on entry into the stroma (Fig. 5.3). The twin arginine residues are recognised by a translocase in the thylakoid membrane, components of which were first recognised from mutants in maize. Three proteins have been identified from mutant maize genes, namely Tha4, hcf106 and cpTatC. Somewhat unusually, the nature of this Tat translocase was identified in plants before an equivalent system was discovered in bacteria, the information from plants aiding the discovery in bacteria. These three proteins all span the thylakoid membrane, but exactly how the three proteins interact to facilitate protein translocation is unclear at present. Two novel characteristics of the Tat thylakoid import system though, are clear. Firstly, transport across the thylakoid membrane is powered by the difference in pH across the membrane (ΔpH), which is maintained by pumping of protons into the thylakoid lumen and the generation of protons in the lumen by the photolysis of water (see Chapter 4). In bright light, there is a difference of three pH units across the membrane equating to a 1000-fold difference in proton concentration. Consequently, no ATP or GTP is required for this mode of lumenal import. It would appear that, as proteins are translocated into the lumen by the Tat mechanism, protons are released back into the stroma and it is estimated that the passage of a single protein into the lumen results in the loss of 30 000 protons from the lumen into the stroma. Secondly, the Tat pathway appears to be able to import proteins into the lumen in a folded state, which stands it apart from the other varied import system in the plastid, which all necessitate unfolded linear proteins. These two mechanisms, Sec and Tat, for import into the thylakoid lumen transport approximately equal numbers of proteins, such that, of those proteins resident in the thylakoid lumen, half are imported by the Sec pathway and half by the Tat pathway. Plastocyanin is imported across the thylakoid membrane into the lumen by the Sec mechanism (Fig. 5.3).

The array of proteins that are targeted to the actual thylakoid membrane itself make up a significant proportion of the protein content of the chloroplasts and consist primarily of the various complexes involved in photosynthetic electron transport including the abundant light harvesting chlorophyll binding proteins (LHC), which bind chlorophyll molecules in the thylakoid membrane. Largely by using reconstitution assays of isolated thylakoid membranes, the way in which several of these proteins become inserted into the thylakoid membrane has been revealed. There are different insertion mechanisms for different groups of proteins (Fig. 5.4). Insertion of LHC, along with a few other proteins, uses a mechanism that involves a stromal signal recognition particle (SRP) and the protein FtsY. Both of these are involved in bacteria in the insertion of proteins into the bacterial plasma membrane, and undoubtedly function in thylakoid biogenesis as a result of the plastid's prokaryotic evolutionary history. LHC proteins contain only a transit peptide sequence to provide transport across the plastid envelope membranes into the stroma and do not have a specific cleavable sequence identifying the thylakoid membrane as a target. Consequently, the stromal SRP particle must recognise an aspect of the LHC protein sequence. The SRP contains two proteins, SRP54 and SRP43, but exactly how the SRP complex interacts with LHC and FtsY is unclear. A further player in this pathway is ALB3, which has homologues in bacteria and mitochondria which function in membrane protein insertion. Evidence for a role of ALB3 comes from examining *Arabidopsis* plants, which carry a mutation in *ALB3*, and in which the thylakoid biogenesis is severely compromised and in which LHC is not inserted into the thylakoid membrane. LHC becomes associated with chlorophyll molecules during thylakoid membrane biogenesis, which is derived from the inner plastid envelope membrane (see Chapter 6).

The vast majority of thylakoid membrane proteins, however, do not use the SRP mechanism for membrane insertion but use a different pathway (Fig. 5.4). Some contain a two-part transit peptide sequence, the first part of which transports them across the envelope membranes into the stroma and the second part of which is called the thylakoid membrane targeting sequence and enables insertion into the thylakoid membrane. However, there is no evidence that this thylakoid membrane targeting sequence is recognised by a specific receptor on the thylakoid or uses ATP or GTP in an insertion mechanism and it is likely that these proteins

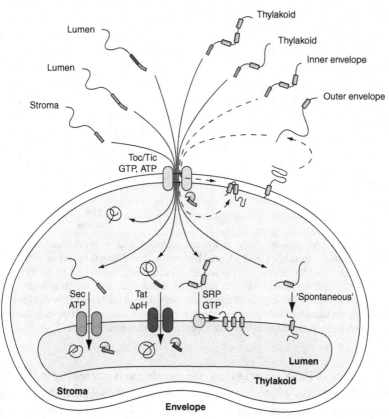

Fig. 5.4. Diagram showing the various routes by which proteins can be targeted to different places within the plastid, either in the stroma, the inner or outer envelope membrane, the thylakoid membrane or the thylakoid membrane lumen. Transport into the thylakoid lumen is by either the Sec or TAT pathways and insertion into the thylakoid membrane is either by SRP or is spontaneous. (Redrawn from Jarvis P, Robinson C. Mechanisms of protein import and routing in chloroplasts. *Current Biology* 14, R1064–R1077. © Elsevier 2004.)

insert spontaneously into the membrane, maybe aided by the presence of the targeting sequence. Once the proteins are inserted into the membrane, a thylakoid processing peptidase enzyme (TPP) cleaves the second transit peptide sequence resulting in the mature protein inserted into the membrane. Proteins inserted in this manner include subunit II of the CF_0 part of the ATP synthase complex, and three proteins involved in photosystem II, PsbW, PsbX and PsbY.

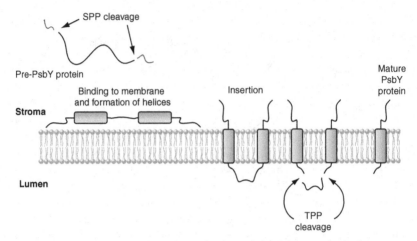

Fig. 5.5. The protein PsbY is nuclear-encoded and is synthesised as a polyprotein containing two identical proteins joined together. After entry into the plastid, a transit peptide sequence is cleaved from either end by SPP and the protein inserts into the thylakoid membrane with two transmembrane α-helices. The thylakoid processing peptidase in the lumen then cleaves out a thylakoid transit peptide, which is present in the region between the two α-helices, resulting in two PsbY proteins in the membrane.

The mode of PsbY insertion into the membrane is interesting in that the precursor protein, which is imported from the cytosol into the stroma, contains two adjoined protein sequences both with cleavable transit peptides enabling transport across the envelope membranes (Fig. 5.5). Once in the stroma, the first transit peptide is removed but the second transit peptide, which intervenes the two mature protein sequences, is only removed after insertion of the whole protein into the thylakoid membrane. The intervening transit peptide sequence is then removed by two cleavage events by the TPP resulting in two adjacent mature PsbY in the thylakoid membrane. Production of two linked proteins in this way is called a polyprotein and is the only known example of such in the chloroplast.

In contrast, the vast majority of thylakoid membrane proteins have no thylakoid targeting sequence, only a transit peptide allowing import across the envelope membranes into the stroma and they appear to insert into the thylakoid membrane by a spontaneous mechanism, which is highly unusual amongst the variety of membrane protein insertion pathways found throughout the cell. Exactly how such a spontaneous insertion pathway would operate is not known, although the highly unsaturated nature of the thylakoid lipid molecules may contribute to the ease of self-insertion.

Evolution of plastid import

Considering the complexity which exists in the import mechanisms and targeting systems within the plastid, it is obvious that substantial evolutionary changes have occurred since the interaction between the primary endosymbiont and its host cell. It is clear that many of the functional components of the various mechanisms that have been described here have counterparts, which still function in modern-day bacteria. Exactly how the transit peptide sequences evolved, upstream of genes once of plastid origin and now resident in the nucleus is unclear, as is the precise evolution of transport complexes on the plastid envelope, which now import these precursor proteins translated in the cytosol. One interesting possibility is that the basis of the import complexes and transit peptide sequences may have evolved from information already present in the primary endosymbiont. Sequencing the genome of the Cyanobacterium *Synechocystis* showed that several Tic Toc proteins show homology to bacterial proteins. Most relevant is Toc75, whose bacterial homologue SynToc75 is involved in export from the bacterial cell and is a member of a general family of export facilitating proteins in bacteria. After the *Toc75* gene moved to the nucleus, the Toc75 protein could have reinserted itself into the plastid envelope but in the reverse orientation, now facilitating import into the plastid rather than export in the Cyanobacterium (Fig. 5.6). It is also possible that information present in the molecules originally exported by SynToc75 in the Cyanobacterium could have been re-used by proteins, which now have to get back into the plastid by a process of exon shuffling. Although this hypothesis appears plausible, it is unclear how other components of the import pathway, particularly those of the Tic complex, have evolved since they show no homologies in Cyanobacterial genomes and probably arose from the evolution of new gene sequences.

The plastid proteome

With the development of technologies enabling the identification of all of the proteins resident inside a given type of plastid, a more global view of

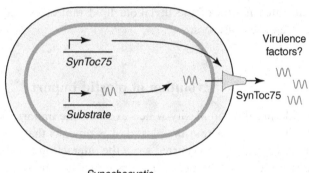

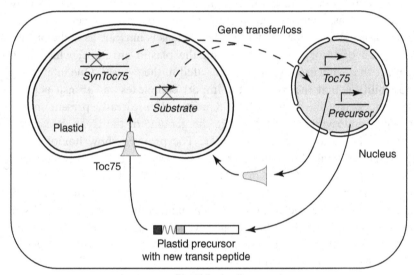

Fig. 5.6. Modern-day transit peptides enabling protein import into plastids may have evolved from proteins enabling secretion of virulence factors in photosynthetic bacteria, such as *Synechocystis*. In the bacteria, a homologue of the Toc75 protein found in modern-day Toc complexes in the plastid envelope is involved in the export of virulence factors from the cell. In the modern-day plant/algal cell, genes for both the Toc75 and the substrate precursor have moved to the nucleus and now the Toc75 resides in the plastid envelope in the opposite orientation directing import of proteins into the plastid. Some of the sequence information in the modern-day transit peptide has come from that original recognised by Toc75 in the substrate protein. (Redrawn from Bruce, B. Chloroplast transit peptides: structure, function and evolution. *Trends in Cell Biology* 10, 440–447. © Elsevier 2000.)

the spectrum of proteins, which are targeted to the plastid, has become available. It is estimated that about 2700 different proteins are present in the chloroplast, a subset of which, with some additions, will be present in other types of plastids. Determining whether a nuclear-encoded protein gets targeted to the plastid by examination of its gene sequence and amino acid sequence should in theory be straightforward if the presence of transit peptide sequences at the N-terminus can be predicted. The most popular software which endeavours to predict the presence of transit peptides within a given sequence is ChloroP, available at http://www. cbs.dtu.dk/services/ChloroP/, although other systems such as Predotar (http://urgi.versailles.inra.fr/predotar/predotar.html), LOCtree (http:// www.predictprotein.org/cgi-bin/var/nair/loctree/query) and MultiLoc (http://www-bs.informatik.uni-tuebingen.de/Services/MultiLoc), which are more general protein targeting prediction packages, are also available. Unfortunately, all of these systems suffer from significant rates of false-positive or false-negative returns, since individual proteins can be tested for precise localisation by attaching a fluorescent protein to them by genetic means and observing directly the localisation of the protein in the cells of a transgenic plant. Thus fluorescent protein fusion experiments remain the best way of finding out the definitive location of an unknown protein in the cell, be it in the plastid, mitochondria or elsewhere. An interesting scenario, mentioned previously in this chapter, is that of dual targeting, in which a single nuclear gene produces a precursor protein capable of being imported into both the plastid and the mitochondria in the same cell. Dual targeting appears to be more widespread than once thought and indeed, of the 24 nuclear-encoded organellar aminoacyl-tRNA synthetases, 17 are dual targeted to both the plastid and the mitochondrion.

Another intriguing fact is that detailed analysis of a large number of the proteins present in isolated chloroplasts, by using tandem mass spectrometry, reveals that around 30% of these proteins, which are encoded by nuclear genes, would not be predicted to be targeted to the chloroplast by the prediction software since they do not have conventional targeting transit peptide sequences. Consequently, it seems possible that other unknown mechanisms function to import nuclear-encoded proteins into plastids. One possible novel mechanism for import into the plastid chloroplast stroma has been discovered by studying the enzyme carbonic anhydrase, which catalyses the conversion of carbon dioxide to bicarbonate ions in the plastid. The carbonic anhydrase protein contains an N-terminal amino acid sequence, which suggests that it is imported into the endoplasmic reticulum and secreted from the cell, in a similar way to its homologues in animal cells. However, a variety of experimental approaches clearly show that the carbonic anhydrase protein resides in the stroma of the chloroplast. Moreover, the protein has sugar groups attached to it by the process of glycosylation, which routinely takes place in the Golgi bodies of the cell. Thus the conclusion from these experiments is that nuclear-encoded proteins can be processed through the secretory system of the cell; the endoplasmic reticulum and the Golgi bodies and are then imported by an unknown mechanism into the chloroplast. It is possible that vesicles derived from the Golgi fuse with the plastid envelope membrane, although this has yet to be proven. Obviously, there is still a large amount to learn about protein targeting to the plastid and the nature of their import mechanisms.

Targeting novel proteins to the plastid

Considering the complexity of the mechanisms that enable the import of proteins into the plastid in plant cells, it turns out that it is relatively easy to fool the system and get novel proteins imported into the plastid that do not normally reside there. Indeed, the targeting of recombinant proteins to the plastid, which are expressed from nuclear transgenes in transgenic plants, has been a major approach in trying to use plants as biofactories to produce large amounts of novel useful recombinant proteins. All that appears to be required to enable entry of a foreign protein into the plastid is the attachment of a transit peptide sequence at the N-terminal of the required protein and the expression of such a transgene in a transgenic plant. The recombinant protein is then imported faithfully through the import mechanisms described previously and the foreign protein accumulates in the stroma (see Chapter 9). Many different soluble proteins have been targeted to the plastid compartment in this way and shown to accumulate in the stroma. Perhaps the most useful protein used to date has been green fluorescent protein (GFP), which is a widely used marker protein used extensively throughout cell biology research. Targeting GFP to the plastid using an additional plastid transit sequence results in GFP accumulating in the stroma. Since transgenic plants carrying such a construct express this transgene in all the cells within a plant, all of the plastids, including those lacking chlorophyll, will fluoresce green when viewed by fluorescence microscopy. This has proved a major benefit to understanding various aspects of plastid biology and morphology, since plastids lacking conventional pigments can now be viewed within the cell. Such a strategy has revealed novel structures of plastid morphology called stromules, which are described in more detail in Chapter 8.

Transport of metabolites and ions across the plastid envelope membrane

So far, we have only considered the import of proteins into the plastid, whereas, in the context of the cell, the plastid imports and exports a wide variety of types of molecules, which are involved in a diverse array of biochemical pathways and general metabolism within the plastid and the cell. The plastid is not only the main site of carbon assimilation within the cell, but also the main site of nitrogen and sulphur metabolism as well as interacting with other cellular organelles in common metabolic pathways, such as photorespiration (see Chapter 7). A proteomic analysis of the plastid envelope membranes reveals that they contain many proteins, the majority of which are likely to be involved in metabolite transport and ion movement, in addition to those involved with protein import complexes. Also, an examination of the fully sequenced genome of *Arabidopsis* has enabled the identification of between 100 and 150 putative transport proteins in the plastid envelope, which are involved in the transport of metabolites and ions across the plastid envelope membrane in either an import or export mode.

The two envelope membranes differ in their ability to transport various metabolites and ions. It was generally considered that the outer envelope membrane is relatively non-selective in its transport properties and that most small molecules can pass across the outer envelope membrane in a relatively non-specific manner, with specificity of import being provided at the inner envelope membrane. The outer envelope membrane contains various proteins, termed Outer Envelope Proteins (OEP); OEP16, OEP21, OEP24 and OEP37, which are named according to their molecular weight and are abundant proteins in the outer envelope membrane. All of them form solute channels by virtue of forming transmembrane domains; OEP16 forms four transmembrane alpha helices and the other three OEP proteins form numerous transmembrane beta sheet structures. Rather than controlling the passage of solutes through the outer envelope membrane in an uncontrolled, unspecified way, experimental evidence suggests that OEP channels are more specific in their transport characteristics. OEP16 appears to be involved with amino acid transport, whereas

OEP21 forms an anion-selective channel, which is regulated by ATP and allows the passage of inorganic phosphate and phosphorylated carbohydrates, such as those exported from the Calvin cycle. OEP 21 is similar to classical porin proteins and appears to be able to transport phosphate, triose phosphate, ATP, dicarboxylate and negatively and positively charged amino acids. This channel may well enable passage of other types of molecules and should be considered a less-specific transport channel. A fourth OEP, OEP37, has been characterised, but the precise nature of its transport abilities are not yet known. Thus, these OEP channels in the outer envelope membrane appear to have much greater specificity over transport of molecules across the outer envelope membrane than was first thought.

Even so, a higher level of specificity of import or export into the plastid is generated by individual transporters present on the inner plastid envelope membrane. It may well turn out that individual transporters exist in this inner membrane specifically for the majority of types of molecules which are required to be imported or exported into the plastid. Another feature of such transport systems, about which little is known, is developmental regulation and the requirement for moving different sorts of molecules in and out of the plastid at different points in its development and in different types of differentiated plastid, in which very different spectra of molecules will be involved. Of those inner envelope transporters characterised in detail, all are encoded by nuclear genes and are inserted into the inner envelope membrane as a result of a transit peptide sequence on the precursor protein, which is eventually cleaved to yield the functional transporter in the inner envelope membrane. This is in contrast to the OEP proteins, which lack any conventional transit peptide sequences. Here, we will consider briefly some of the characterised inner envelope membrane specific transporters.

An important family of transporters in the plastid envelope are the phosphate translocator proteins, which export various phosphorylated sugar molecules out of the plastid in exchange for importing phosphate ions back into the plastid. The triose phosphate/phosphate transporter (TPT) is abundant in the plastid envelope, constituting about 10% of the total protein on the inner envelope membrane. The 36 kD protein has several transmembrane domains and probably functions in the membrane as a dimer. TPT exports triose phosphate molecules in the form of glyceraldehyde 3-phosphate (GAP) or dihydroxyacetone phosphate (DHAP) from the stroma into the cytosol, where they are used primarily to synthesise sucrose. These triose phosphates are generated by carbon dioxide fixation in the photosynthetic Calvin cycle in the stroma of the chloroplast, and these exported molecules represent the profit resulting from photosynthetic carbon fixation (see Chapter 4). In exchange for triose phosphate export, TPT imports phosphate ions in a strict 1:1 ratio with the exported molecule, thereby ensuring that the stroma does not run short of phosphate and it becomes limiting during photosynthesis. A second transporter of this family, PPT, translocates phosphoenol pyruvate (PEP) across the plastid envelope membrane into the cytosol, an important requirement in the mesophyll cells of C_4 plants in which carbon dioxide is fixed by the carboxylation of PEP by the enzyme PEP carboxylase in the cytosol. PEP export in these cells is required to ensure that cytosolic carboxylation can continue unabated (see Chapter 8). In non-green plastids, PPT is involved in the import of PEP from the cytosol into the plastid stroma where PEP is involved in the synthesis of aromatic compounds via the shikimate pathway. Two other members of this family of transporters transport simple phosphorylated sugars across the plastid envelope in exchange for phosphate molecules. Glucose 6-phosphate is transported by the glucose 6-phosphate/phosphate translocator (GPT) and xyulose 5-phosphate is transported by the xylulose 5-phosphate/ phosphate translocator (XPT). GPT function is important in non-green plastids where it imports glucose 6-phosphate, which is the primary source of carbon for these plastids, which are non-photosynthetic and heterotrophic.

Two different transporters are present in the plastid envelope, which transport dicarboxylate molecules. DiT1 imports 2-oxoglutarate into the plastids and at the same time exports malate into the cytoplasm in a counter-exchange mode. A related transporter, DiT2, imports malate from the cytoplasm and at the same time exports glutamate. Thus, the overall effect of the concerted action of these two transporters is to import 2-oxoglutarate and export glutamate with no net movement of malate. These two transporters function in the pathway of photorespiration in which ammonia is assimilated in the plastid as a result of 2-oxoglutarate import and the action of the enzymes glutamine synthetase (GS) and ferrodoxin-dependent glutamate 2-oxoglutarate aminotransferase (Fd-GOGAT) (see Chapter 7).

Two other transporters are involved in the export of sugars from the plastid, which result from the degradation of starch stored at the end of the light period in the chloroplasts as a result of photosynthetic carbon assimilation. pGlcT is a 43 kD protein on the inner envelope membrane with 12 membrane spanning helices and exports glucose from the chloroplast into the cytsol. A hexose kinase enzyme located on the outer envelope of the chloroplast immediately converts the glucose to glucose 6-phosphate, a key molecule in the synthesis of sucrose in the cytsol (see Chapter 7). A different transporter, Mex1, exports maltose as a result of nocturnal starch degradation. Mex1 resides in the inner envelope membrane and has nine membrane spanning helices. In the endosperm of cereals, the synthesis of starch in the plastid uses ADP-glucose molecules, which are synthesised in the cytsol and require import into the plastid. The transporter that imports ADP-glucose was originally identified from a mutant of maize, called *brittle1* (Bt1) and it imports ADP-glucose in these cells into the plastid, probably exporting AMP in a counter-exchange mode. Bt1 proteins are present in the plastid envelope of other plastid types including chloroplasts, but their precise role in these photosynthetic plastids is unclear.

The inner envelope membrane also contains a transporter, which specifically transports the adenylate molecules ADP and ATP across the envelope membrane. This transporter is crucial in the metabolic function of non-green plastids, which are involved in the synthesis of various storage molecules such as starch or fatty acids and require ATP from the cytosol in order to drive this synthesis. Thus the AATP1 transporter imports an ATP molecule in counter-exchange for the export of an ADP molecule.

There are many other classes of metabolite molecules that are required to be imported or exported by the plastid, such as pyruvate, amino acids, glycolate and glycerate to name a few, but the nature of the transporters that move them is unknown. Analysis of those putative transport proteins revealed by genomic sequence and membrane proteomics of the plastid envelope is likely to reveal considerably more details about the range of transporters present in the envelope membranes of different types of plastids, and the nature of the molecules which they transport.

The control of the movement of simple and complex ions into and out of the plastid is crucial if the plastid is to maintain its synthetic capacity and homeostasis of the stromal compartment. Many different ions are implicated in a wide range of plastid processes, and plastids take up inorganic cations including H^+, K^+, Na^+, Ca^{2+}, Cu^{2+}, Mn^{2+}, Mg^{2+}, Fe^{2+} and Zn^{2+}, as well as a range of inorganic anions including PO_4^{3-}, SO_4^{2-} and NO_2^-. Not only is the plastid the major site of sulphur and nitrogen metabolism in the cell, it also requires a variety of metal ions, which form functional groups within essential proteins within the plastid or are co-factors associated with enzymes. Although many potential transporter proteins have been identified by electrophysiological and membrane transport studies, and potential candidates for specific transporters have been identified in proteomic analysis of the plastid envelope membrane, detailed characterisation of individual transporters and analysis of what they transport and how they function has only proceeded significantly since the millennium and it is still unknown how most ions get across the inner envelope membrane. Most progress has been achieved by using gene sequence information from sequenced genomes and by making comparisons with ion transporters in other biological systems such as prokaryotes or animal cells or by identifying mutants that show impaired uptake of a specific ion into the plastid and show a deleterious phenotype as a result.

A variety of metal anions are used in many different processes within the plastid, especially in redox reactions associated with electron transport on the thylakoid membrane. Foremost amongst these is iron (Fe^{2+}), which is used extensively in thylakoid membrane complexes. The plastid contains the cell's major functional store of iron, which is sequestered by a complex of ferritin proteins within the plastid in the form of a ferric-oxy-hydroxide. Ferritin proteins are encoded by a family of nuclear genes and imported into the plastid, where they form a complex of 24 subunits with an iron core. Iron is moved from this store when it is required, in order to synthesise several different iron-containing complexes within the plastid including Fe–S proteins, ferredoxin and cytochromes on the thylakoid membrane or Fe-protoporphyrin-IX in the haem biosynthetic pathway. Iron is transported into the plastid from the cytosol in the ferrous form (Fe^{2+}) by the PIC1 transporter, which resides in the inner envelope membrane and shows homology to permease-like proteins found in Cyanobacteria.

Magnesium is an essential nutrient in plants, having a major role in chlorophyll biosynthesis, with a magnesium ion in the porphyrin head group of the pigment. Magnesium deficiency in plants is typically observed as a yellowing of tissues and the inability to synthesise sufficient chlorophyll. The concentration of magnesium ions (Mg^{2+}) are also important in the control of Calvin cycle enzymes (see Chapter 4) and, during light-induced photosynthesis, magnesium ions move from the thylakoid lumen into the stroma, thereby activating Calvin cycle enzymes. The way in which Mg^{2+} ions are imported into the plastid appears to be via an inner envelope transporter termed AtMRS2-11, a member of the AtMRS2 gene family in *Arabidopsis*, which shows homology to the CorA family of magnesium transporter proteins found in Eukaryotes.

Copper ions (Cu^{2+}) are also vitally important in plastid function in two major ways. Firstly, copper is a co-factor associated with plastocyanin, the small mobile carrier found in the thylakoid lumen, which moves electrons from the cytochrome b_6f complex and donates them to PSI. Secondly, copper together with zinc are co-factors with the stromal enzyme superoxide dismutase (Cu/ZnSOD), an enzyme involved in scavenging reactive oxygen species formed on the thylakoid membrane during light-induced electron transfer, especially around the oxygen evolving complex associated with PSII (see Chapter 4). Thus copper is required both in the stroma and in the thylakoid lumen in the chloroplast. It is transported to these sites by two P-type ATPases, which transport Cu^{2+} across membranes, hydrolysing ATP to ADP in the process. PAA1 resides in the inner envelope membrane and moves copper ions from the cytosol into the stroma, where it can associate with SOD enzymes. A second transporter, PAA2, is present in the thylakoid membrane and transports copper ions into the thylakoid lumen, where it can associate with plastocyanin protein. Plants carrying a mutation in PAA1 still manage to import copper into the plastid, suggesting that a second mechanism may exist. This was confirmed by identifying a second copper transporter protein residing in the plastid envelope, another P-type ATPase termed HMA1. These two plastid envelope copper transporters may have different physiological roles for copper import in differing environmental conditions in the cell.

Manganese is a third metal ion which has a crucial role in plastid function, specifically as a co-factor in the oxygen evolving complex on the thylakoid lumenal side of photosytem II. Thus, like copper, manganese ions (Mn^{2+}) need to be imported into the chloroplast and moved across the thylakoid membrane. The precise nature of these transporters is not known, although a member of the Nramp family of transporter proteins, which are general divalent cation transporters coupled with proton exchange, characterised in other eukaryotic systems, would be a likely candidate for at least part of this manganese import pathway.

Calcium ions (Ca^{2+}) are a widely used signalling molecule in eukaryotic cells and calcium ions undoubtedly play a significant role in plastid function, not least in that calcium ions are required in synthesis of the oxygen-evolving complex associated with PSII and calcium ions are associated with controlling activation of an array of plastid enzymes. Knowledge of exactly how calcium ions cross the inner envelope membrane remains elusive. There is evidence, however, for the existence of a Ca^{2+}/H^+ antiporter mechanism in the thylakoid membrane which imports Ca^{2+} into the thylakoid membrane whilst exporting protons into the stroma, although the exact nature of this protein is not known.

The control of hydrogen ion concentration in the stroma and thylakoid lumen is a key aspect of plastid function and, during light-driven photosynthesis, the pH of the stroma is maintained at an alkaline pH, around 8, whilst the thylakoid lumen is at a pH around 6 or less, because of proton accumulation in the lumen. The alkaline pH of the stroma in the light is crucial for the activation of several Calvin cycle enzymes to ensure a high photosynthetic rate and is maintained by control of proton movement across the chloroplast envelope by a $Na^+(K^+)/H^+$ exchange protein, which can export H^+ into the cytosol and simultaneously import Na^+ or K^+ from the cytosol into the stroma, thereby maintaining an alkaline pH in the stroma, and facilitating Na^+ and K^+ entry into the plastid. Such antiporter proteins of this type exist in many different cellular membranes and the plastid member of this family, termed AtCHX23, functions to maintain pH homeostasis between the chloroplast stroma and the cytsol.

The plastid is also the primary site of nitrogen and sulphur assimilation in the cell (see Chapter 7) and thus there is a requirement for transport of sulphur- and nitrogen-containing ions across the plastid envelope. Nitrogen is imported into the plastid in the form of nitrite (NO_2^-) or ammonium (NH_4^+) ions. Nitrite ions have been shown to be imported rapidly into the plastid by a nitrite transporter, CsNitr1-L, in the plastid envelope, which is a member of the proton-dependent oligopeptide family (POT) of transporter proteins. Sulphur is imported into the plastid in the form of sulphate (SO_4^{2-}) ions (see Chapter 7) by a sulphate transporter in the inner plastid envelope, which is likely to be similar to the Sulp sulphate permease transporter which imports sulphate ions into algal chloroplasts, although the situation in higher plants has yet to be fully clarified and the sulphate import protein characterised fully.

It is clear from our consideration of the plastid envelope that it is a highly dynamic system, which moves very significant numbers of molecules of a very diverse type and size across the double membrane into the plastid stroma. In addition, it also exports many molecules in bulk out into the cytosol. Many aspects of its biology and transport properties have yet to be characterised properly, but the next decade will likely see major progress in this area.

6

The development of the chloroplast

During the growth and development of a seedling plant, the various types of plastids which are found in different cell types develop from those proplastids found in the cells of the shoot and root meristems. These proplastids provide the source for all plastids in all the different types of cells within the plant. The way in which proplastids develop into different plastid types in different types of cells is poorly understood, but by far the best-researched plastid developmental pathway is that of the proplastid developing into a mature chloroplast. This developmental pathway occurs primarily in the mesophyll cells of leaves, resulting in large populations of mature green chloroplasts in these cells, although chloroplast development can also occur to varying degrees in all other green tissues of a plant. The development of a mature chloroplast in a leaf mesophyll cell or in other green tissues requires a huge synthesis of proteins, lipids and metabolites, the vast majority of which are imported from the cytosol into the plastid. In particular, an extensive array of proteins encoded in the nucleus and translated on cytosolic ribosomes are required to be imported using the import mechanisms described in Chapter 5.

Moreover, a major feature of this chloroplast developmental pathway is coordination. Not only is coordination required between the expression of nuclear genes encoding proteins destined for the chloroplast and the expression of those genes encoded on the chloroplast's own genome, but also coordination is required between the expression of these nuclear genes and the developmental status of the chloroplast, as well as its functional status in terms of photosynthetic activity and the degree to which the chloroplast is stressed by environmental conditions. Since the photosynthetic protein complexes that reside on the thylakoid membrane are composed of proteins encoded in the two different genomes, nuclear and plastid, it is important for the plastid to coordinate their synthesis such that complexes can be built efficiently and effectively. Once the basic features of the chloroplast such as the extensive thylakoid membranes and these photosynthetic protein complexes have been constructed, there is also the ongoing regulation of coordination between plastid and nucleus, since continual import of proteins and metabolites is required to maintain the chloroplast's dynamic metabolism and to service the turnover of proteins. Consequently such coordination is an ongoing, dynamic affair and not just required in the early stages of plastid construction.

Since the original endosymbiotic events leading to the earliest plastids inside eukaryotic cells, various complex signalling systems have evolved to ensure that chloroplast development is tightly regulated within the plant cell and that the nucleus has effective control over chloroplast construction and function.

A crucial feature of the development of the chloroplast from the proplastid is the requirement for light (Fig. 6.1). Without light being present in the circadian cycle, chloroplasts fail to develop correctly and in seedlings grown in the dark, etioplasts develop instead of chloroplasts (see Chapter 2). In the natural world, the formation of etioplasts probably only occurs in exceptional circumstances where growing plants are completely deprived of light for a length of time. However, the plastids found in the cotyledonary cells of young seedlings are etioplast-like in that they have developed during seed development as far as they are able in relative darkness within the developing seed and they wait for a light signal as the seed germinates in order to finish their conversion to functional green chloroplasts. This happens in those cotyledons which emerge above the ground during hypogeal germination, whereas in those seeds having epigeal germination, where the cotyledons remain below the ground in plants such as peas, these cotyledonary plastids never fully become chloroplasts. In normal leaf development, however, the shoot apical meristem is illuminated by sunlight and this enables the proplastids that are present in those cells in leaf primordia, which are destined to become mesophyll cells, to embark upon a pathway of chloroplast differentiation. In addition, the plastids in these cells must replicate extensively in order that the mature mesophyll cells contain large populations of individual mature chloroplasts, to maximise photosynthetic efficiency and light capture. This process of plastid division will be considered later (see Chapter 8).

At the earliest stage of chloroplast development, we must consider a small population of proplastids, around 20, in an undifferentiated cell in a leaf primordia, which has just been initiated on the dome of the shoot

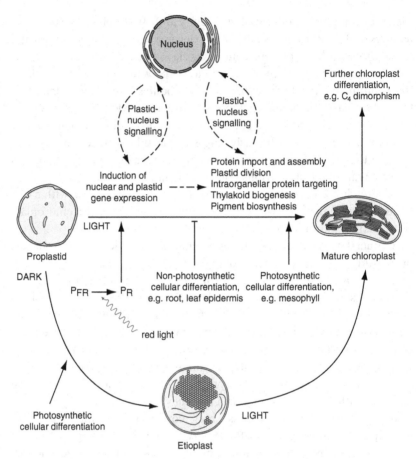

Fig. 6.1. The development of the chloroplast from the progenitor proplastids in higher plants requires light for the process to proceed. Without light, etioplasts are formed, which can redifferentiate into chloroplasts when illumination recommences. The photoreceptor phytochrome (P_{FR}, P_R) is central to the control of this process by light (see Fig. 6.8). The process of chloroplast development also requires ongoing communication between the chloroplast and the nucleus to ensure efficient coordination between the two organelles in terms of gene expression, protein production and assembly of macrocomplexes, such as those on the thylakoid membrane. (Redrawn from Pyke K, Waters M (2005). Plastid development and differentiation. In *Plastids* (ed. Møller SG). *Annual Plant Reviews* 13 with permission from Wiley-Blackwell Publishing.)

apical meristem. Individual proplastids normally contain some small amounts of thylakoid membrane, little if any chlorophyll, a DNA nucleoid and their membranes will contain low levels of various transporter proteins. Once the commitment to become a mesophyll cell has been initiated, a train of events is put in place, which involves synthesis of all the components needed to build a mature chloroplast, including the accumulation of genetic information and translational capacity in the form of ribosomes. Firstly, we will consider how the extensive thylakoid membrane network is built.

Making the thylakoid membrane network

The thylakoid membrane is perhaps the most obvious feature of the chloroplast's structure and is easily observed in electron micrographs of sectioned chloroplasts as an extensive membrane system, which extends throughout the chloroplast (Fig. 6.2). The thylakoid membrane is best imagined as a large single bag, which is squashed to form a thin sheet with space inside: the thylakoid lumen. This sheet of membrane is then convoluted into a highly complex architecture, although it remains a single contiguous compartment within the stroma. Two distinct components of thylakoid architecture are apparent in sectioned chloroplasts. The granal stacks can contain up to 15–20 layers of thylakoid membrane and in extreme cases over 100. Different granal stacks are linked together by single layers of stromal thylakoid lamellae, which enter the grana at differing heights in the stack. Consequently, individual layers of thylakoid membrane can either exist as granal lamellae or stromal lamellae (Fig. 6.2). Although the two-dimensional appearance of the thylakoid membrane can be clearly seen in electron micrographs, moving from these two-dimensional images to predict the fully three-dimensional structure of the membrane architecture has proved difficult to determine. A striking model for thylakoid structure that has been developed from careful analysis of electron micrographs is that of stacks of fused granal lamellae forming a right-handed helical structure where the lamellae appear like blades of a fan and where stromal lamellae join stacks together at alternating levels within a stack (Fig. 6.3).

The average chloroplast contains significant amounts of thylakoid membrane to the extent that the total volume of thylakoid membrane and its internal lumen takes up around 20% of the chloroplast's volume and, in an average leaf, it is estimated that the total area of thylakoid membrane is in the order of several hundred square metres. Consequently, the packing efficiency of membrane on which light energy can be captured greatly exceeds the simple area of a leaf. It is estimated that the area to volume ratio of the thylakoid in relation to plastid volume is

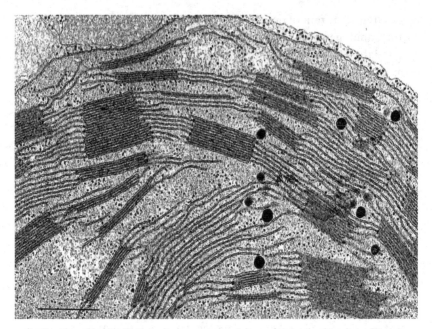

Fig. 6.2. A beautiful electron micrograph image of a transverse section through the thylakoid membrane network in a chloroplast. The granal stacks of membrane vary greatly in number and are interlinked by stromal lamellae, which join the granal stacks at alternate levels in the stack. It is important to realise that this image is a rather artificial two-dimensional image of a section through a three-dimensional system and if the plane of sectioning was altered, the image would look very different! Bar = 0.2 μm. (Image courtesy of http://botit.botany.wisc.edu/images/130/Plant_Cell/Electron_Micrographs/ chloroplastgrana.)

between 7×10^7 and 5×10^8 m^{-1}, which compares with 10^6 m^{-1} for a solid sphere of radius 3 μm, an approximation of a chloroplast.

It is important to realise that granal stacks of the thylakoid membrane are not permanent structures and they only form as a result of attractive interactions between the surfaces of two adjacent sheets of thylakoid membrane, mainly in the form of electrostatic interaction between proteins on the membrane surface and van der Waals' forces between adjacent molecules on appressed membrane surfaces. Thylakoid grana isolated from chloroplasts *in vivo* can be routinely unstacked and restacked by altering their ionic environment. Another factor that contributes to granal stacking is the presence of complexes of chlorophyll and protein that are associated with photosystem I and photosystem II and which act as antennae systems for light capture (see Chapter 4).

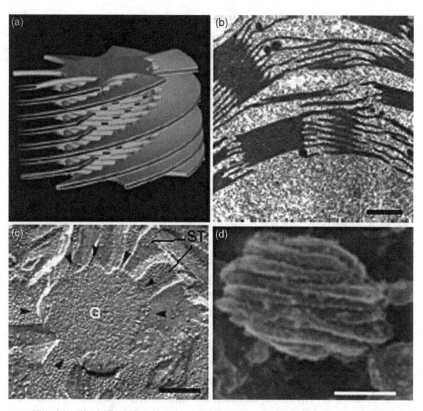

Fig. 6.3. The helical arrangement of thylakoid membranes in a granal stack is computer modelled in (a), which reflects what is seen by different types of electron microscopy in images (b)–(d). The computer-generated model was constructed from serial sections, such as that shown in (b), through a granal stack. Some of the stromal lamellae have been omitted, so one can see how the stromal lamellae join the helical core of the granal stack. In image (c), the granal stack has been frozen and fractured and viewed by scanning electron microscopy. In this image, the core of the granal stack (G) has fractured at right angles to its height and one is looking at a flat sheet of granal core membrane, with the peripheral stromal lamellae (ST) arranged in a helical manner. In image (d), a granal stack has been isolated from prepared thylakoid membranes by sonication, leaving the granal core intact but missing the peripheral stromal lamellae. Bars = 0.2 μm. Mustardy L, Garab G. Granum revisited. A three-dimensional model – where things fall into place. (*Trends in Plant Science* 8, 117–122. © Elsevier 2003.)

Considering the importance of the thylakoid membrane for photosynthesis and chloroplast function, the knowledge of how this complex architecture is put together is scant. Firstly, there needs to be extensive synthesis of lipids in order to build such large amounts of membrane and, subsequently, there needs to be a degree of coordination in synthesising such complex architecture. The lipid synthesising enzymes, which generate the thylakoid membrane, are associated with the plastid envelope and the lipid composition of the thylakoid membrane is similar to that of the inner plastid envelope membrane. The major groups of lipids found in the thylakoid membrane are the neutral galactolipids mono- and digalactosyldiacylglycerol (MGDG and DGDG). These constitute around 85% of the total thylakoid lipids, the rest being composed of the negatively charged lipids sulphquinovosyldiacylglycerol (SL) and phosphatidylglycerol (PG), which account for 14%, with phosphatidylinositol (PI) making up 1% of total lipids. Although the envelope membranes are the site of much of this lipid synthesis, the diacylglycerol in these lipids can be derived from different sources, either from the plastid or from the endoplasmic reticulum (ER). This latter route necessitates that fatty acids have to be exported from the plastid to the endoplasmic reticulum and then returned to the plastid by a mechanism, the details of which are unclear, although a permease-like protein, TGD1, which resides in the outer envelope protein, is involved in the import of lipid precursors from the ER. There is flexibility between these two pathways since different species can use predominantly one or the other and plants with a mutation in the plastid-specific biochemistry simply revert to using the ER pathway. Several different MGDG synthase enzymes are present on the outer and inner envelope membranes and DGD1, which synthesises DGDG from MGDG molecules, is associated with the outer envelope membrane. A major part of thylakoid membrane synthesis, which is poorly understood, is how lipids made on the plastid envelope get transported and incorporated into the thylakoid membrane in the stroma. At present, most evidence suggests that vesicles bud off from the inner envelope membrane and are transported through the stroma to the growing thylakoid, with which they fuse subsequently. In addition, occasional connections between the thylakoid membrane and the inner envelope membrane are observed in mature chloroplasts (Fig. 6.4). Vesicles can be observed budding off from the inner envelope and in situations where plants are chilled, these vesicles can accumulate in significant numbers close to the inner envelope membrane (Fig. 6.4). Exactly how these vesicles bud and fuse with the growing thylakoid membrane, what is in their membranes and what cargo they carry, are questions that remain to be answered although a few proteins have been implicated in this general process. ARF1 and Sar1 are nuclear-encoded GTPases located in the chloroplast, which are homologous to components of the cytosolic ER-Golgi trafficking system and appear to

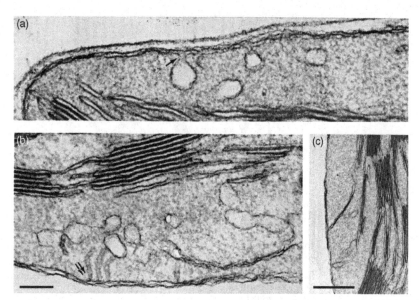

Fig. 6.4. Electron micrographs of thylakoid biogenesis derived from vesicles, which bud off of the inner plastid envelope membrane and give rise to thylakoid membranes in the stroma. In image (a), vesicles can be seen budding from the inner envelope membrane and in image (b), linear pieces of membrane are formed linking the inner envelope to vesicles (arrowed). In image (c), a clear connection between the inner envelope membrane and the thylakoid membranes can be seen. These images are from leaf discs of tobacco cooled to 12 °C, which slows down vesicle trafficking allowing them to be observed more easily. In normal grown leaves at higher temperatures, observation of vesicle trafficking from the envelope to the thylakoid is much rarer. Bar = 0.2 μm O (Morre JD, Sellden G, Sundqvist C, Sandelius AS (1991). *Plant Physiology* 97, 1558–1564. © American Society of Plant Biologists and reproduced with permission.)

function in plastid vesicle trafficking, along with two other proteins, VIPP1 and Thf1. When either VIPP1 or Thf1 are mutated, vesicle production from the inner envelope membrane is not observed and thylakoid biogenesis is perturbed. VIPP1 forms a high molecular weight complex on the inner envelope membrane, which may be associated with vesicle production.

Chlorophyll biosynthesis

An important aspect of the light control of chloroplast development in higher plants is the synthesis of the green pigment chlorophyll. Chlorophyll is a tetrapyrolle molecule and is synthesised entirely within the plastid organelle from the amino acid glutamate (Fig. 6.5) by a complex

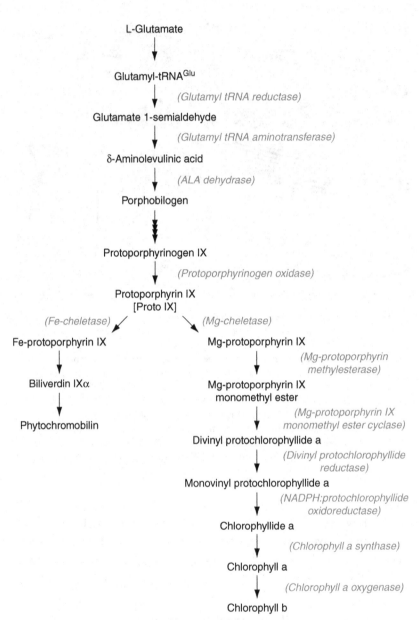

Fig. 6.5. The synthesis of chlorophyll takes place in the chloroplast and shares a common pathway at the beginning with that of haem. The starting molecule for the pathway is the amino acid glutamate and the pathways diverge by the insertion of magnesium into protoporphryin IX rather than iron, which occurs in the haem pathway. The major control point for light control of chlorophyll synthesis in Angiosperms is the enzyme NADPH: protochlorophyllide oxidoreductase, which in the light generates chlorophyllide a. The conversion of chlorophyll a into chlorophyll b is carried out by

pathway involving 15 different enzymes, the early parts of which are common to the synthesis of other tetrapyrrole molecules such as haem. All these enzymes are encoded by nuclear genes and the enzymes are all synthesised on cytosolic ribosomes and imported into the plastid. The enzymes which catalyse the early steps of the pathway (Fig. 6.5) are soluble and function in the stroma, but the enzymes that catalyse the final few steps to generate chlorophyll are associated with the inner plastid envelope membrane and it is here that chlorophyll molecules are synthesised and complexed with proteins. The control of chlorophyll synthesis by light is manifested in higher plants by the enzyme that catalyses the penultimate step in the pathway, the conversion of protochlorophyllide to chlorophyllide a. Protochlorophyllide is a colourless molecule, whereas chlorophyllide a is green and thus this is the first step in chlorophyll biosynthesis that generates a coloured molecule. The enzyme that carries out this step is called proto-chlorophyllide oxidoreductase and in higher plants its activity is dependent on the presence of light. Consequently, in higher plants grown in the dark, the enzyme is unable to carry out the reaction and significant amounts of the protochlorophyllide reductase enzyme and its substrate accumulate in large arrays of prolamellar bodies in the etioplasts, which form under such dark conditions (see Chapter 2). When light is eventually perceived, the reaction commences and green chlorophyllide a is synthesised. This light-dependent reaction in higher plants is a fundamental light control step for chloroplast development. However, in the vast majority of instances where chloroplasts develop from proplastids in normally grown seedlings or leaves, light is present and there is no intermediate etioplast step but a sequential development from proplastids to chloroplasts. In stark contrast, a version of the protochlorophyllide oxidoreductase enzyme found in Gymnosperms, lower plants and photosynthetic bacteria is light independent and functions in the dark. Consequently, seedlings of Gymnosperms such as pines are green when grown in the dark in contrast to a higher plant, such as *Arabidopsis*, in which dark grown seedlings are pale and lack green pigment. The final part of the chlorophyll biosynthetic pathway converts chlorophyllide a into chlorophyll a, a reaction catalysed by the enzyme chlorophyll a synthase. Subsequently, a modified form of chlorophyll a, chlorophyll b, is generated by the action of the enzyme chlorophyll a oxidase (Fig. 6.6).

Caption for Fig. 6.5. (cont.)
chlorophyll a oxygenase, which resides in the envelope membranes, as do all of the enzymes in the chlorophyll-synthesising branch of the pathway. The relevant enzymes are shown for most of the steps in the pathway. Several of the *gun* mutants of *Arabidopsis* carry mutations in genes encoding different enzymes of the chlorophyll biosynthetic pathway.

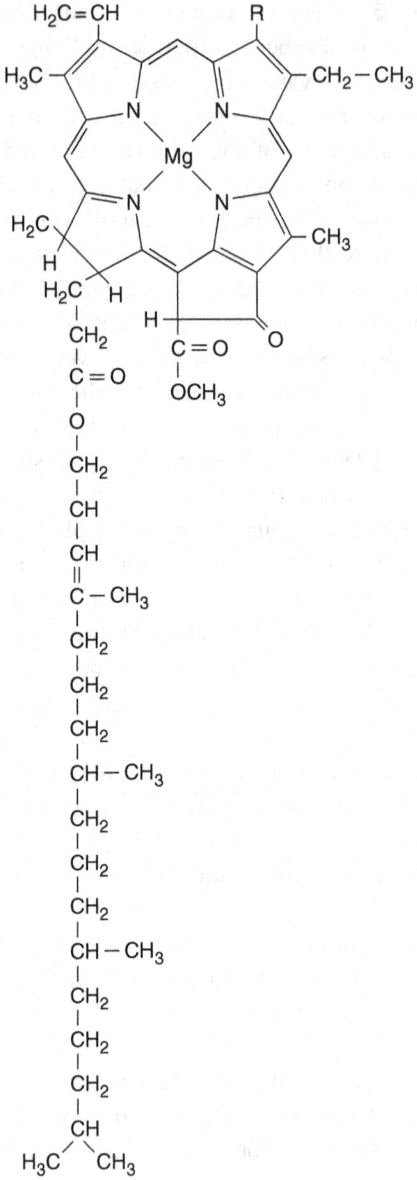

Fig. 6.6. The chlorophyll molecule contains a porphyrin head group containing a magnesium atom and a long phytol tail. The production of chlorophyll b from chlorophyll a by the enzyme chlorophyll a oxidase involves the conversion of the R group from CH_3 in chlorophyll a to CHO in chlorophyll b.

Not only is chlorophyll required for light harvesting and energy transduction in photosynthesis, but it has also become clear that it is fundamental in enabling the construction of the photosynthetic complexes. Upon synthesis, chlorophyll molecules are complexed with LHC proteins and form into light-harvesting antennae complexes which are associated with the two photosystems on the thylakoid membrane, PSI and PSII. Whether the vesicles that generate the thylakoid membrane already have thylakoid proteins as cargo or inserted in the vesicle membrane is an interesting question, which also remains to be answered clearly. However, what is clear is that chlorophyll is essential for the complete assembly of the photosynthetic apparatus on the thylakoid membrane since, if chlorophyll synthesis is perturbed, such as in the dark, then assembly of the photosynthetic apparatus is incomplete. LHCII proteins are encoded by nuclear genes and are imported into the plastid through the Tic/Toc complexes by virtue of a transit peptide sequence. There is evidence from experiments examining the synthesis of LHCII that newly imported LHCII molecules require binding to chlorophyll to stabilise them and that the LHCII apoprotein inserts and assembles with chlorophyll in the inner plastid envelope membrane first, before being transported to thylakoids, either through vesicles or by new thylakoid budding off of the inner envelope membrane directly (Fig. 6.7). As discussed in Chapter 5, the SRP complex and the ALB3 protein are also critical to the process by which imported LHC molecules are inserted into the membrane. In addition to chlorophyll binding to stabilse LHC, secondary pigment molecules, neoxanthin and violoxanthin, are also required (Fig. 6.7).

Proplastid to chloroplast development: the role of light

As we have seen, light is a major factor in enabling differentiation of chloroplasts from proplastids as a result of the stimulation of chlorophyll synthesis. Light also controls several other aspects of chloroplast development, including the expression of nuclear-encoded genes targeted to the plastid, the expression of plastid-encoded genes on the plastid genome and the translation of the resulting mRNA molecules by plastid ribosomes. In order for light to have a controlling effect on this plastid differentiation pathway, light must be received by the cell or the chloroplast and its energy absorbed and the result of that event must be perceived by a signalling system, which then moderates cellular or plastid processes. Enabling of the chloroplast differentiation process to occur in

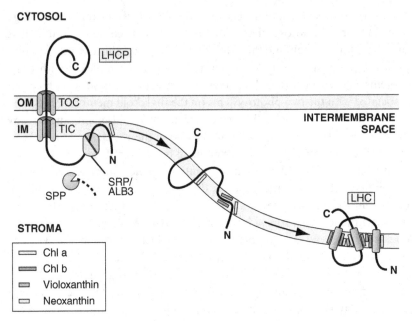

Fig. 6.7. The import of light-harvesting chlorophyll binding protein (LHC) occurs as a precursor (LHCP), which is imported via the Toc and Tic complexes. The N-terminus transit peptide sequence is cleaved by the stromal processing peptidase (SPP) and the LHC moves into the inner envelope membrane with the aid of the chloroplast SRP complex and the ALB3 protein. The terminal region is inserted in the inner membrane whilst the rest of the linear protein is being imported. On insertion into the membrane, chlorophyll a and chlorophyll b molecules bind to the membrane spanning parts of the LHC protein, which eventually results in the fully processed LHC protein, with three membrane spanning domains, being bound with several chlorophyll molecules and also neoxanthin and violoxanthin molecules. These chlorophyll protein complexes then migrate into the main thylakoid membrane either by vesicle trafficking from the inner envelope or possibly through regions of membrane, which are contiguous between the two membrane systems.

developing seedlings is part of a wider process of plant development that requires light and is called photomorphogenesis. This process involves not only chloroplast differentiation but also seedling hypocotyl elongation and leaf expansion and is mainly driven by the absorption of light by the photoreceptor phytochrome. Phytochrome is a cytosolic photoreceptor which is interconvertible between two forms P_R and P_{FR}. P_R absorbs red light and converts to an active form P_{FR}, which can enter the nucleus and interact with transcription factors to moderate nuclear gene

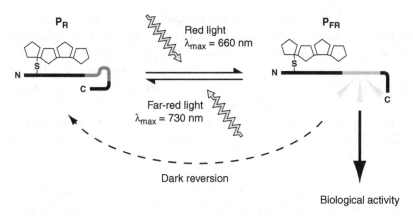

Fig. 6.8. The photoreceptor phytochrome exists in cells in two forms, inter-converted by the absorption of red and far red light. The inactive form P_R absorbs red light and changes its conformation to generate P_{FR}, which is biologically active. P_{FR} reverts to P_R either by absorbing far red light or by a slow reversion during the dark. Thus the relative balance of red and far red light in the photoenvironment of the cell will dictate the equilibrium levels of P_R and P_{FR} and the extent of biological activity.

expression. Absorption of far red light by P_{FR} inactivates P_{FR} by con-verting it back to P_R (Fig. 6.8). In the dark, activated P_{FR} reverts to P_R or is degraded. Thus, the presence of the red light component in sunlight and the ratio of red : far red light will induce activation of phytochrome and dictate the amount of active phytochrome present. In addition to this type of phytochrome, a second type exists, encoded by the *PHYA* gene and inactive P_R is only synthesised in the dark. This type of phyto-chrome is extraordinarily sensitive to light and is chiefly responsible for the first photomorphogenesis events. A large number of the nuclear-encoded proteins, which are destined for the chloroplast, have their gene expression controlled by phytochrome such that their expression is light stimulated, requiring red light and, as a result, little expression occurs in the dark. Two transcription factors, PIF1 and PIF3, bind to a specific G-box sequence CACGTG in the promoters of light-regulated nuclear genes, including those destined for the plastid. Their binding to these promoter elements appears to be either positively or negatively regulated by the interaction with nuclear-localised active P_{FR} molecules. The inter-action of PIF3 and activated P_{FR} and its interaction with G-box sequences induces the expression of a significant number of genes encod-ing proteins destined for the plastid, along with many other genes involved in other cellular processes. In contrast, the PIF1 transcription factor plays a specific role in the control of chlorophyll biosynthesis by negatively regulating the production of the protochlorophyllide in the chlorophyll biosynthetic pathway. Thus, in the absence of activated phytochrome P_{FR} molecules in the nucleus, PIF1 inhibits transcription of the nuclear genes encoding enzymes involved in the chlorophyll biosynthetic pathway but on illumination, the repression is removed by binding with active P_{FR} and chlorophyll biosynthesis can proceed. One reason why such a system may have evolved is to prevent the accumulation of large amounts of intermediate molecules in the chlorophyll biosynthetic pathway, which may be toxic within the chloroplast. Since protochlorophyllide reductase also requires light to carry out its enzymatic function within the chlorophyll biosynthetic pathway, it is clear that chlorophyll biosynthesis is tightly controlled by light in at least two different ways.

Light and plastid gene expression

As was mentioned in Chapter 3, one way in which the expression of plastid genes is controlled by light is by the light-stimulated expression of the nuclear genes encoding sigma factors, which are required by the plastid-encoded RNA polymerase to enable transcription. A good example of both the simplicity and complexity of the light control of plastid gene expression is the plastid gene *psbD*, which encodes the D2 protein, which functions in the reaction centre of photosystem II on the thylakoid membrane. The *psbD* promoter on the plastid genome is responsive to blue light, by way of a blue light receptor, cryptochrome. Cryptochromes are encoded by two nuclear genes *CRY1* and *CRY2*, and the cryptochrome proteins are localised in the nucleus, where they absorb blue light and become activated. Once activated, cryptochromes can enhance the expression of specific nuclear genes, thus making their expression light-controlled. The expression of the nuclear gene for a specific sigma factor, *AtSig5* which is involved in plastid transcription, is rapidly upregulated by activated cryptochrome and the resulting AtSig5 protein is imported into the plastid, where it facilitates the binding of the plastid-encoded polymerase to the *psbD* promoter on the plastid genome, thereby stimulating *psbD* gene expression in response to blue light. The expression of other plastid genes encoding reaction centre proteins, such as *psbA*, encoding the D1 protein of the reaction centre of photosysytem II, and *psaA* and *psaB*, which encode the two major proteins of the reaction centre of photosystem I, is also controlled by light but through a different control system: that of redox sensing.

The redox control of plastid processes appears to play a central role in the dynamic control of plastid function, at the level of photochemistry and light absorption, as well as at the level of the control of plastid gene expression. Experiments suggest that the redox control has effects at many different stages in the process from initiation of transcription, through mRNA processing, mRNA transcript stability, ribosome loading and translation initiation and subsequent protein production and that different plastid-encoded genes are affected differently. Two major sources of redox control have been shown in the chloroplast. The redox state of the plastoquinone pool is measured as the proportion of plasto-quinone molecules which have accepted electrons (reduced), compared with those which have not (oxidised). A major point of redox control is the expression of the plastid genes encoding the proteins at the core of PSII and PSI; *psbA* and *psbB* for PSII and *psaA*, *psaB* and *psaC* for PSI. Control of their expression in relation to the redox state of plasto-quninone acts as a feedback mechanism controlling the stoichiometry of the amounts of PSII and PSI on the thylakoid membrane. Thus, in a situation where the plastoquinone pool is mostly reduced and electrons need to be passed efficiently through to PSI, the redox signal induces activation of the PSI core protein genes and represses the transcription of PSII core protein genes. As a result, this redox sensing maintains a balance in the stoichiometry of PSII and PSI to maintain optimal electron flow. Such a mechanism runs on a longer-term scale with a greater lag time than the similar redox-controlled mechanisms of state II-state I transitions of LHC movement between the photosystems described in Chapter 4.

Another mechanism by which the redox state controls plastid gene expression is through the function of a plastid transcription kinase (PTK). This enzyme is nuclear-encoded and has the ability to phosphoryl-ate polypeptides associated with the plastid transcription machinery, including subunits of the plastid-encoded RNA polymerase and sigma transcription factors. Activity of PTK is controlled by reduction of cyst-eine residues within its structure by glutathione. Reduced glutathione is generated in the plastid by the enzyme glutathione reductase, which uses NADPH and is a key part of the antioxidant response to oxidative stress in the plastid. However, in this context, the production of reduced glutathione provides a link between electron transport and the produc-tion of NADPH and the activity of PTK. Phosphorylation of plastid transcriptional components appears to reduce transcription activity and appears to affect both photosynthetic and non-photosynthetic plastid gene transcription. Although such light-induced controls of plastid transcription by different mechanisms are important in the control of protein production, it is generally considered that post-transcriptional processes and aspects which control the translation of plastid-encoded proteins are more important features in the control of plastid protein production than the rate of transcription itself.

Assembly of the thylakoid protein complexes

It is obvious that the control of the expression of genes encoding proteins, which reside in the thylakoid membrane, is complex and that there are several different regulatory mechanisms involved in both the control of gene expression and the control of translation of the resulting mRNA molecule. Once proteins have been translated, the thylakoid complexes need to be assembled. There are two distinct aspects to this process. Firstly, there is the initial phase of synthesis during the early development of the chloroplast, which requires an enormous amount of protein and lipid synthesis to construct a mature thylakoid membrane. Once photosynthetic electron transport is functional, a second aspect is that of the control of protein turnover in the thylakoid membrane complexes and the removal of damaged proteins and their replacement with newly synthesised proteins.

Since the photosynthetic complexes are composed of a significant number of different protein subunits, there is a requirement that the supply of proteins to be assembled is consistent and that all required proteins are available at a given time. The mode of assembly is best understood for the photosystem II complex, which is built in a stepwise manner. A key player in this assembly process is cytochrome b_{559}, which can accumulate in the thylakoid membrane in the absence of other PSII proteins. Cytochrome b_{559} interacts with newly synthesised D2 proteins, encoded by the plastid *psbD* gene, and forms a precomplex. This precomplex then binds the D1 protein, the product of the plastid gene *psbA*, and forms the essential core of PSII, consisting of a D1/D2 heterodimer. With the addition of *psbI* and *psbW* gene products, a complex is formed that can bind all essential redox co-factors of PSII and is the smallest PSII complex that can carry out light-induced charge separation. Subsequently, light-harvesting antennae subunits associate with this complex followed by the gene products of plastid genes *psbB* and *psbC*, namely CP47 and CP43 are added, which then allows the binding of the proteins involved in the oxygen-evolving complex on the lumenal side of the complex, which are the products of nuclear genes *psbO*, *psbP* and *psbQ*. By analysing mutants in which individual subunits are missing or their expression is perturbed, it is clear that there is interaction between different subunits in controlling each other's production, particularly in the control of plastid-encoded subunits. For example, if the *psbA* gene is mutated on the chloroplast genome and thus no D1 protein is made, there is a coordinated decrease in the synthesis of CP47 protein but not of D2 protein. Thus, the presence of a given subunit is required for the continued synthesis of another subunit in the same complex, which is plastid-encoded. This type of mechanism is called control by epinasty of synthesis (CES) and is a means by which efficient coordinated production of different subunits within a complex can be facilitated. A good example of this type of assembly process is shown by the construction of the cytochrome b_6f complex, which contains cytochrome f, cytochrome b_6 and subunit IV. Translation of cytochrome f mRNA declines dramatically if unassembled cytochrome f proteins, which are not associated with their partners cytochrome b_6 or subunit IV, are present for a period. This mechanism involves free cytochrome f molecules preventing translation initiation of their own mRNA molecules by the binding of the C-terminal domain of the cytochrome f protein to the mRNA ribosome complex. This type of mechanism also occurs during the translation of D1 and CP47 proteins, where unassembled proteins inhibit their own translation. Thus by such complex interactions, mature PSII complexes can be assembled efficiently, and large populations of unassembled subunits are not present in the plastid for significant periods of time.

The ongoing control of protein turnover in thylakoid complexes is critical to maintaining functional complexes, which can carry out photosynthetic electron transfer. The chloroplast cannot afford to have significant numbers of photosynthetic complexes non-functional for long periods of time. Thus, damage to proteins within the complexes has to be monitored and damaged proteins have to be excised and newly synthesised ones inserted to replace them. This is particularly true of core reaction centre proteins, which are easily damaged by excessive light absorption and the resulting photochemical reactions, in the process of photoinhibition. The best understood case is the turnover of the D1 protein at the core of PSII and the repair of PSII complexes (Fig. 6.9). In this D1 repair cycle, damaged PSII complexes are phosphorylated and

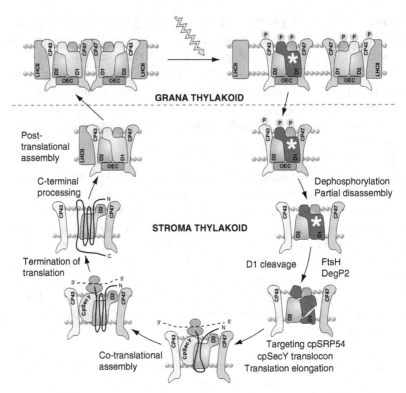

Fig. 6.9. The repair cycle that replaces damaged D1 proteins in photosystem II complexes involves phosphorylation and movement of damaged complexes to the stromal lamellae of the thylakoid membrane, partial disassembly and degradation of damaged D1 proteins by proteases, co-translational insertion of newly synthesised D1 protein in association with the Sec translocon and post-translational processing and assembly followed by migration back to the stacked granal membrane. (Redrawn from Baena-Gonzalez E, Aro E-M (2002). Biogenesis, assembly and turnover of photosystem II units. *Philosophical Transactions of the Royal Society London* 357, 1451–1460 with permission from The Royal Society.)

move from granal lamellae to stromal lamellae, where they undergo partial disassembly to allow protease access to the damaged D1 protein. A key feature of this cycle is the coordination between removal of degraded D1 proteins and synthesis and reinsertion of newly synthesised molecules. A complex of the *psbA* mRNA bound to the ribosome associates with the thylakoid membrane, binding to the Sec translocon channel in the thylakoid membrane, a feature that was discussed in Chapter 4 and enables the import of proteins into the thylakoid lumen. In this case, it acts as an anchor for the *psbA* mRNA ribosome complex, which then guides the insertion of the D1 protein into the PSII complexes in a co-translational manner. Elongation of the D1 protein at this point requires functional photosynthetic electron transfer, one result of which is that D1 repair can only happen in the light and not in the dark. On completion of translation, the PSII complex reforms and the D1 C-terminal sequence which extends into the thylakoid lumen is cleaved and the newly repaired PSII complexes migrate back to the granal lamellae (Fig. 6.9). It is important to realise that this repair cycle is a constant dynamic process in the illuminated chloroplast, which increases its turnover dramatically in high light, where conditions of photoinhibition may occur and where damage to D1 increases. D1 is the most highly turned-over protein in the thylakoid membrane, with an average half-life of about 20 minutes. Thus, the synthesis of new D1 proteins in the illuminated chloroplast represents a major flow of molecules.

Signalling from the plastid to the nucleus

During the evolutionary development of the plastid as a distinct organelle in plant cells, many changes have occurred. Most significantly, the movement of genes from the plastid genome to the nucleus has led to a requirement for the plastid and the nucleus to communicate in order to coordinate expression of those genes now resident in the nucleus with the state of development of the plastid. In addition, the plastid, and the chloroplast in particular, are highly dynamic in terms of ongoing bio-chemical metabolism and photosynthetic activity, and the nucleus needs to be aware of the extent of such activities in order to modify its gene expression to optimise plastid performance. Since the chloroplast is largely controlled by gene products from the nucleus, the bulk of the communication flows from the nucleus to the plastid and is termed anterograde signalling. In contrast, the plastid needs to communicate with the nucleus as a monitor of development within the plastid, of the state of photosynthetic activity in the plastid and the potential for stress conditions arising within the plastid. Such signalling from the plastid to the nucleus is termed retrograde signalling. Evidence that the chloroplast communicates with the nucleus in this way comes from several different types of experimental observations in which plastid development is perturbed and which results in a reduction in the level of gene expression in the nucleus of genes encoding plastid-targeted proteins. For instance, white sectors can be induced in leaves, in which mutation in a nuclear gene prevents the accumulation of carotenoids in the plastids. In such cells, there is a down-regulation of nuclear genes encoding chloroplast-targeted proteins such as LHC and the small subunit of the carbon fixation enzyme RUBISCO. This reflects a general observation that the absence of functional chloroplasts within a cell causes a down-regulation in gene expression amongst groups of nuclear genes encoding plastid-targeted proteins. Such a situation can be artificially induced by inhibitors, which target various plastid processes such as norflurazon, which inhibits carotenoid biosynthesis in the plastid, lincomycin, which inhibits plastid translation, or nalidixic acid which inhibits plastid DNA synthesis.

The conclusions from such experiments suggested that retrograde signalling from the plastid to the nucleus was real and thus more extensive searches for the plastid processes that might be involved in initiating such signalling pathways were made. The isolation of mutants of *Arabidopsis* in which such a signalling pathway has been perturbed was important in revealing components of such a pathway. These *gun* (*Genomes UNcoupled*) mutants fail to alter their nuclear gene expression of plastid-targeted genes when normal chloroplast function is perturbed and hence reveal genes which play a role in such a retrograde signalling process. Identification of the genes encoded by *GUN2*, *GUN3*, *GUN4* and *GUN5* show them all to encode enzymes which function in the early parts of the chlorophyll and haem biosynthetic pathway in the chloroplast (Fig. 6.5) and highlight the fact that, in each mutant, there is minimal accumulation of the chlorophyll intermediate Mg-protoporphyrin IX, which is on the chlorophyll branch of the tetrapyrrole biosynthetic pathway (Fig. 6.5). Thus, an increased level of Mg-protoporphyrin IX in the chloroplast appears to initiate a signalling pathway that results in altered gene expression in the nucleus, especially that of LHC genes. The reason for such a system is that if Mg-protoporphyrin IX accumulates significantly in the plastid, then complete biosynthesis of chlorophyll must be perturbed and consequently the requirement for the nuclear-encoded chlorophyll binding protein *LHC* is reduced.

A second source of retrograde plastid signals appears to emanate from the expression of plastid genes on the plastid genome. Evidence for this pathway came from the analysis of seedlings treated with a plastid translation inhibitor such as chloramphenicol. In such seedlings, plastid proteins are not made but a down-regulation in nuclear gene expression for plastid-targeted proteins is also observed. This type of signalling occurs early in seedling development and is independent of light. Consequently, it would appear that, during the early development of the plastid, correct gene expression and translation on plastid ribosomes are required to initiate a supply of proteins encoded by nuclear genes.

A third source of retrograde signals results from redox signalling as a monitor of the state of photosynthetic electron transport on the thylakoid membrane of the chloroplast. As was discussed in Chapter 3, the redox state of the plastoquninone pool can modify expression of genes encoded on the plastid genome, but it can also change patterns of nuclear gene expression. A variety of conditions can induce changes in the redox state of components of the photosynthetic electron transport apparatus in the chloroplast, including moving low-light grown plants to high light, changes in temperature or sugar starvation or the blocking of electron transport with herbicides. A key component, which has been shown to be involved in this retrograde signalling, is the redox state of the plastoquinone pool. Thus the proportion of plastoquinone molecules which have accepted electrons (reduced) compared with those which have not (oxidised) appears to be important in this process. Analysing the resulting changes in nuclear gene expression when the redox state of the photosynthetic electron transport components alters shows that a large number of nuclear genes are directly regulated by this process.

The way in which these three retrograde plastid signals: Mg-protoporphyrin IX levels, redox state and plastid gene expression, manifest an effect on gene expression in the nucleus appears to be via a protein called GUN1, which was identified from analysis of the *gun1 Arabidopsis* mutant. The GUN1 protein is targeted to the plastid and is a pentatrico-peptide repeat protein, a class of proteins which bind to RNA and DNA. Most significantly, all three retrograde signalling pathways appear to signal to the nucleus via the GUN1 pathway, thereby generating a common signal from all three systems (Fig. 6.10). Since GUN1 binds DNA, it seems likely that it works by affecting in some way the expression of plastid genes. Exactly how this works and the exact nature of the physical signal that communicates across the cytoplasm to the nucleus is not known. However, when the signal gets to the nucleus, it manifests its effect on nuclear gene expression via the transcription factor ABI4, which binds to genes encoding plastid-targeted proteins and prevents their expression. Thus a system is initiated which prevents further synthesis of plastid proteins when their requirement is reduced because of particular conditions within the chloroplast.

In addition, the presence of reactive oxygen species in the plastid also contributes to a retrograde signal. Reactive oxygen species such as singlet

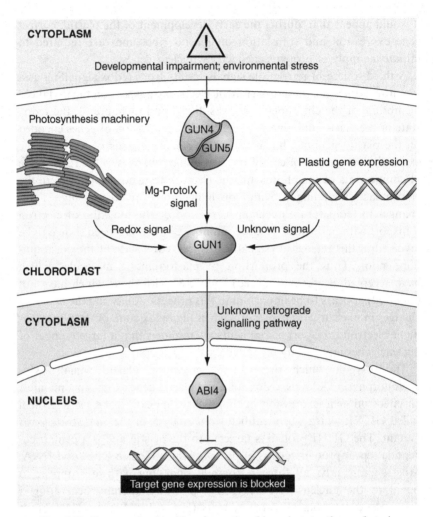

Fig. 6.10. Retrograde signalling from the chloroplast to the nucleus is manifested through the action of GUN proteins. Three different signals: Mg-protoporphyrin IX accumulation, plastid gene expression and redox signalling from the electron transport chain coalesce their effect through the GUN1 protein, which effects a signal transduction pathway, which changes patterns of gene expression in the nucleus, mostly a shut-down in the transcription of genes encoding plastid-targeted genes. Exactly how GUN1 signals from the chloroplast across the cytoplasm to the transcription factor ABI4 in the nucleus is unclear. (From Zhang D-P (2007). Signalling to the nucleus with a loaded gun. *Science* 316, 700–701. Redrawn with permission from the American Association for the Advancement of Science.)

oxygen $^1O_2$ can be generated by the photochemical reactions on the thylakoid membrane, particularly under conditions of high light but are normally detoxified by antioxidant systems in the plastid such as the enzyme superoxide dismutase. Levels of singlet oxygen, together with hydrogen peroxide (H_2O_2), which is formed by the action of peroxide enzymes, both appear able to initiate signalling to the nucleus. The reason why the nucleus requires information about the redox state of the photosynthetic electron transport pathway or the abundance of these reactive oxygen species is so that sizes and functionality of the chloroplast antennae complexes associated with the two photosystems can be modified by changes in nuclear gene expression, particularly that of *LHC* genes, or by expressing genes encoding stress response proteins or enzymes, which synthesise antioxidant molecules such as carotenoids. Thus such a signalling system would constantly monitor the state of the chloroplast in this regard and modify gene expression accordingly. Exactly how these reactive oxygen species generate a signal to the nucleus is unclear, but a chloroplast-targeted protein called EXECUTER1 has been shown to function somehow in this signalling process.

7

Plastid metabolism

As we have seen in preceding chapters, the plastid plays a major role in the plant cell in carrying out photosynthesis and enabling the plant to increase in biomass as a result of carbon dioxide assimilation. In addition, the plastid also plays a crucial role in carrying out a variety of metabolic processes, which give rise to a myriad of different molecules, which are used both inside the plastid or are exported into the cytosol. As a result, the original endosymbiont that was taken up by an early eukaryotic cell has become so important to the functioning of the modern-day plant cell that it is generally considered that cells lacking plastids are non-functional and not viable. Thus the plastid is a prerequisite for plant cell function.

Furthermore, it is important to realise that a significant proportion of the plastids resident in the cells of a higher plant are not photosynthetic chloroplasts but other non-photosynthetic types such as leucoplasts, amyloplasts, root plastids or chromoplasts and that these plastids carry out many critical parts of cellular metabolism in these non-photosynthetic tissues which are essential to cell function. These plastids are generally termed non-green plastids.

Research on a diverse range of plastid metabolism has generated a huge amount of information about plastid biochemical pathways, the details of many of which are beyond the scope of this book. In this chapter, therefore, we will consider an overview of the main biochemical pathways that occur in plastids, especially those associated with the chloroplast.

Sucrose and starch synthesis

The central metabolic cycle in the chloroplast is the Calvin cycle, also called the reductive pentose phosphate pathway (RPPP) and which was considered in some detail in Chapter 4. This cycle enables the fixation of carbon dioxide by the enzyme RUBISCO and the subsequent metabolism of the resulting products to regenerate substrate for the fixation reaction and to generate carbon-containing molecules for metabolic synthesis in the plastid and in the cell as a whole. The major export from this Calvin cycle are triose phosphate molecules in the form of dihydroxyacetone phosphate and glyceraldehyde 3-phosphate, which are exported from the plastid by the triose phosphate transporter in the plastid envelope in strict counter-exchange for the import of phosphate ions (Fig. 7.1). In the cytosol, triose phosphate is used to synthesise fructose 6-phosphate and UDP-glucose, which combine to eventually form sucrose, a cytosolic pathway that is largely regulated under the control of fructose 2-6 bisphosphate (Fig. 7.1). Sucrose thus formed in the cytosol can be transported around the plant, primarily from the leaves to the roots or developing seeds, by the phloem transport system and constitutes the major energy and carbon store used by the plant to enable growth and the synthesis of longer-term storage molecules such as starch and lipids. In some plants, sucrose itself is a storage molecule, particularly in the two major sugar-producing crops, sugar cane (*Saccharum officinarum*) and sugar beet (*Beta vulgaris*). Sucrose is also a signalling molecule, which has been implicated in the control of a wide variety of molecular and physiological processes in plants. There is some evidence that sucrose is also found within the plastid itself, although what its role might be within the plastid is unclear. However, a more abundant storage material, which is only synthesised within the plastid, is starch. Starch is synthesised in two different scenarios inside plastids; either as a transient short-term storage molecule made during the illuminated part of the diurnal cycle to store photosynthate derived from carbon fixation or as a long-term storage molecule within amyloplasts in specific storage tissues, such as tubers or seeds (see Chapter 2). In these two situations, the source of carbon for starch synthesis is different. In the chloroplast, some of the excess carbon which has been fixed during photosynthesis and which has not been exported to the cytosol as triose phosphate is used for the biosynthesis of starch, in the form of fructose 6-phosphate, which is an intermediate of the Calvin cycle (Fig. 7.1). The enzyme hexose phosphate isomerase converts fructose 6-phosphate to glucose 6-phosphate, which is then converted into glucose 1-phosphate by the enzyme phophoglucomutase. The first committed step in starch biosynthesis is then the formation of ADP-glucose by ADP-glucose pyrophosphorylase. ADP-glucose is the building block with which the starch polymer is made via the enzyme starch synthase. There are multiple

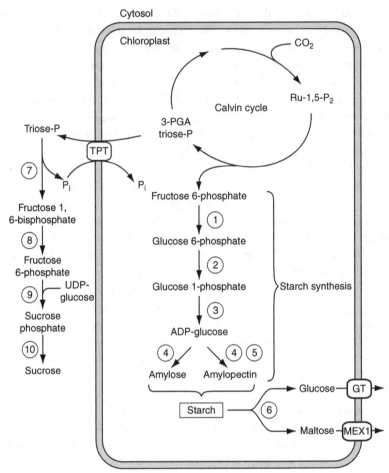

Fig. 7.1. Metabolic pathways that synthesise sucrose and starch are a result of carbon fixation in the chloroplast. Some of the fixed carbon, which is not exported from the plastid as triose phosphate via the triose phosphate transporter (TPT) is removed from the Calvin cycle as fructose 6-phosphate. A series of enzyme-driven reactions then synthesises ADP-glucose inside the plastid, which is the primary building block for starch synthesis. The enzymes in the pathway are (1) hexose phosphate isomerase, (2) phosphoglucomutase, (3) ADP-glucose pyrophosphorylase, (4) starch synthase, (5) starch branching enzyme. Starch is broken down largely at night by amylase enzymes (6) to produce maltose and glucose, which are transported from the plastid by their respective transporters in the plastid envelope membrane. Sucrose synthesis is initiated in the cytosol by the conversion of exported triose phosphate into fructose 1-6 bisphosphate by the enzyme fructose 1-6-bisphosphate aldolase (7), which is then converted to fructose 6-phosphate by fructose bisphosphatase (8). The enzyme sucrose phosphate synthase then generates sucrose phosphate (9), which in turn is converted to sucrose by sucrose phosphatase (10).

isoforms of starch synthase enzymes, some of which are involved in building amylose polymers and others involved in building amylopectin polymers, these being the two types of starch polymer. The amylose-forming starch synthases are bound to the starch granule, whereas the amylopectin-forming starch synthases are more generally distributed between the stroma and starch granules. The branched structure of the amylopectin molecule is formed by starch-branching enzymes (Fig. 7.1).

In starch synthesis in non-photosynthetic cells, such as in amyloplasts in storage tissues or in columella cells in the root cap (see Chapter 2), the carbon source for starch synthesis has to come from imported molecules which have been translocated from photosynthetic parts of the plant. In such cases, the main source of carbon that is imported into the amyloplast is glucose 6-phosphate, which enters via the glucose 6-phosphate/Pi antiport transporter in the plastid envelope, in which phosphate ions are exported in exchange for glucose 6-phosphate import (Fig. 7.2). Glucose 6-phosphate is then converted to glucose 1-phosphate, and starch is synthesised using the same set of enzymes as used in chloroplasts. This starch synthesis requires ATP, which has to be transported from the cytosol by an ATP/ADP transporter in the plastid envelope. In starch synthesis in the endosperm of developing cereal grains, starch can be synthesised in the same way but there is an additional pathway which occurs in that ADP-glucose can be synthesised in the cytosol first and is then imported into the amyloplast where starch synthase converts it to starch (Fig. 7.2).

The breakdown of starch stored in chloroplasts or amyloplasts requires a controlled degradation of the starch grains, which form significant structures within the plastid (see Chapter 2). Starch breakdown in the transient starch grains, made in the chloroplast during the day, occurs during the following dark night period. There appears to be little starch breakdown during the day in these chloroplast starch grains, such that degradation commences at night and the degradation mechanisms are probably inhibited by light. In amyloplasts in storage tissues, starch grain degradation is initiated by regrowth of the plant or by germination of the seed, which contains amyloplast-filled endosperm. In cereal grains such as barley and wheat, starch breakdown is initiated during early germination by the gibberellin-induced activation of genes encoding amylase enzymes in cells of the aleurone layer, which surround the endosperm tissues. Various types of amylase enzymes have the ability to break down different parts of the complex starch molecule. α-amylase cleaves the α-(1-4)-glucosyl bonds, which join sequential glucose molecules into chains,

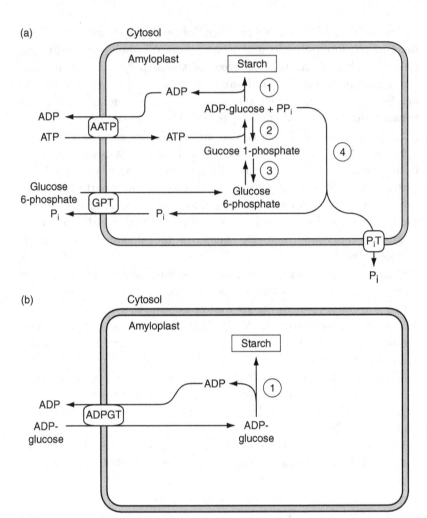

Fig. 7.2. (a) Starch synthesis in heterotrophic non-green plastids uses imported glucose 6-phosphate as a source of carbon. ATP is imported to drive reactions, in counter-exchange for ADP. Pyrophosphate (PPi), which is produced during starch synthesis, is broken down to phosphate and exported. Enzymes are (1) starch synthase, (2) ADP-glucose pyrophosphorylase, (3) phosphoglucomutase, (4) pyrophosphatase. (b) Although the scheme shown in (a) may happen in cereal endosperm amyloplasts, they also have the capability of importing ADP-glucose itself and synthesising starch from it using (1) starch synthase. (Redrawn from Tetlow IJ, Rawsthorne S, Raines C, Emes MJ (2005). Plastid metabolic pathways. In *Plastids* (ed. Møller SG). *Annual Plant Reviews* 13 with permission of Wiley-Blackwell Publishing.)

resulting in various branched malto-oligosaccharides, which are then acted on by β-amylase and starch phosphorylase to produce maltose and glucose 1-phosphate, respectively. Maltose has been shown to be the major molecule exported from the chloroplast at night by a specific transporter termed MEX1 in the plastid envelope (see Chapter 4). Although the majority of the enzymes that function in the starch degradation pathway have been characterised, exactly how they function to degrade an intact starch grain remains to be established.

Although the main biochemical pathways through which starch is synthesised and degraded are known, exactly how all these enzymes interact and are regulated is less clear, but there is a high level of control and protein–protein interaction between enzymes. Analysis of starch-synthesising enzymes isolated from starch grains show that they form a complex, the activity of which is dependent upon its phosphorylation status by a plastidial protein kinase. In particular, types of starch-branching enzymes and starch phosphorylase enzymes together form a starch grain-associated complex when they are phosphorylated, which becomes unassembled when the enzymes become dephosphorylated. Another point of control for starch-metabolising enzymes comes from the surprising discovery that the amyloplast contains the entire ferredoxin/thioredoxin system, which has been characterised in chloroplasts and functions during light-induced photosynthesis (see Chapter 4), namely ferredoxin, ferredoxin–thioredoxin reductase and thioredoxin. The reduction of ferredoxin to activate this thioredoxin system in amyloplasts in heterotrophic cells, such as developing endosperm, is carried out not by light-driven electron transport, but by NADPH. NADPH is generated by metabolism within the amyloplast itself, as a result of sucrose and hexoses import from the phloem via long-distance transport from photosynthetic leaves in a distant part of the plant. Many different amyloplast enzymes, including some of those associated with starch metabolism are thus under redox control as manifested by thioredoxin, and this mechanism appears to be a way of linking the biosynthetic activities of the amyloplast with the photosynthetic capacity of the leaves, a relationship normally referred to as source-sink communication.

The oxidative pentose phosphate pathway

In photosynthetic chloroplasts, the major source of NADPH used to drive different aspects of metabolism is derived from photosynthetic

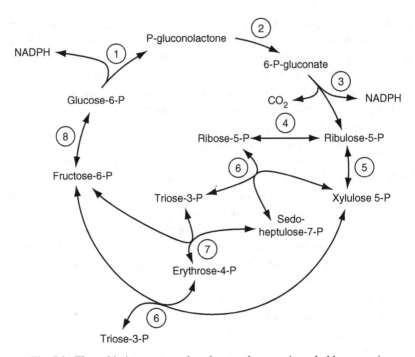

Fig. 7.3. The oxidative pentose phosphate pathway as it probably occurs in plastids, producing NADPH for use in other metabolic pathways. Enzymes are (1) glucose-6-phosphate 1-dehydrogenase, (2) 6-phosphogluconolactonase, (3) 6-phosphogluconate dehydrogenase, (4) ribose-5-phosphate isomerase, (5) ribulose-5-phosphate-3-epimerase, (6) transketolase, (7) transaldolase, (8) glucose-6-phosphate isomerase. (Redrawn from Kruger NJ, von Schaewen. The oxidative pentose phosphate pathway: structure and organisation. *Current Opinion in Plant Biology* 6, 236–246. © Elsevier 2003.)

electron transport resulting from light absorption on the thylakoid membrane. However in non-green plastids, which are heterotrophic rather than autotrophic, NADPH has to be generated in a different way, since these plastid types are not photosynthetic. In heterotrophic plastids such as leucoplasts, root plastids and amyloplasts, NADPH is generated by the oxidative pentose phosphate pathway, termed **OPPP** (Fig. 7.3), which also generates several intermediate molecules used in other biosynthetic pathways. The first part of the pathway is termed the oxidative part and utilizes imported glucose-6-phosphate to generate ribulose-5-phosphate in a three step process involving the enzymes glucose-6-phosphate dehydrogenase, 6-phosphogluconolactonase and 6-phosphogluconate dehydrogenase, two steps of which generates an NADPH molecule (Fig. 7.3). The non-oxidative part of the pathway involves the interconversions of different phosphorylated sugars and generates xyulose-6-phosphate, ribose-5-phosphate, sedoheptulose-7-phosphate and erythrose-4-phosphate, several of which are used in subsequent biosynthetic pathways within the plastid. The NADPH-generating oxidative part of the pathway also occurs in the cytosol, with the two sets of enzymes involved being encoded by two discrete sets of genes, one set being the plastidial form and one set being the cytosolic form. In contrast, the non-oxidative part appears to be restricted to the plastid compartment, although this may change according to developmental status and environmental factors. Many of the enzymes involved in this latter process have genes encoding distinct isoforms, which may function in either the plastid or the cytosol.

Nitrogen assimilation and amino acid biosynthesis

Plastids play a central role in the assimilation of nitrogen in plants and the biosynthesis of many of the 20 amino acids that are the building blocks of proteins. In roots, the plant takes up nitrogen from the soil in the form of either nitrate (NO_3^-) ions or ammonium (NH_4^+) ions and imports them into the cytosol by their respective transporters in the cell's plasma membrane (Fig. 7.4). Once in the cytosol, nitrate ions are reduced to nitrite ions by the cytosolic enzyme nitrate reductase, which carries out the following reaction:

$$NO_3^- + NADPH + H^+ \rightarrow NO_2^- + NADP^+ + H_2O.$$

Nitrate ions within the cell are partitioned between a metabolically active pool in the cytosol and a stored pool of nitrate ions in the vacuole. Nitrate reductase is a highly regulated enzyme, the activity of which is controlled by phosphorylation and binding to 14-3-3 cytosolic proteins, as well as by many environmental factors such as light, which plays a role in up-regulating the synthesis of the enzyme. The nitrite ions produced by this reaction are then rapidly imported into the plastid, where all subsequent reactions of primary nitrogen assimilation and amino acid biosynthesis occur, in both photosynthetic and non-photosynthetic plastids. Nitrite ions pass across the plastid envelope membranes via the plastid nitrite transporter (see Chapter 5). Once in the plastid stroma, nitrite ions are converted to ammonium ions (NH_4^+) by the enzyme nitrite reductase, which is a nuclear-encoded protein imported into the plastid and whose

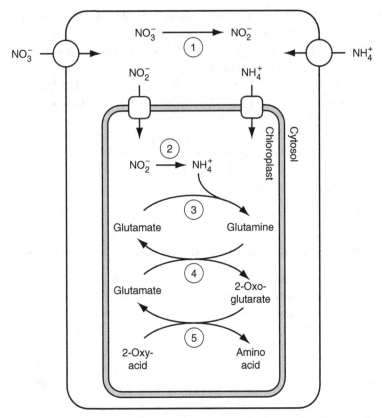

Fig. 7.4. Assimilation of nitrate and ammonium ions occurs in the plastid after conversion of nitrate to nitrite, which is imported into the plastid along with ammonium ions. The so-called GS/GOGAT cycle gives rise to glutamate and other amino acids produced by transaminase enzymes. Enzymes are (1) nitrate reductase, (2) nitrite reductase, (3) glutamine synthetase, (4) glutamine:oxoglutarate aminotransferase (GOGAT), (5) amino transferases.

gene expression is dramatically up-regulated by nitrate ions. Nitrite reductase carries out the reduction of nitrite to ammonium ions by the reaction:

$$NO_2^- + H^+ + Ferredoxin_{reduced} \rightarrow NH_4^+ + Ferredoxin_{oxidised} + H_2O.$$

In photosynthetic tissues containing chloroplasts, the source of electrons for this reaction is from reduced ferredoxin associated with PSI on the thylakoid membrane. In non-green plastids, such as those in roots, the source of electrons is less clear. The ammonium ions thus formed are toxic and are rapidly metabolised into the amino acid glutamate by two enzymes; glutamine synthetase and glutamate synthase, also called glutamine:oxoglutarate aminotransferase or GOGAT. These two enzymes form the so-called GS/GOGAT cycle in plastids (Fig. 7.4). Ammonium ions which have come from other sources can also be assimilated in this way; either those taken up directly from the soil or those arising from the breakdown of nitrogenous compounds in the cytosol, which are part of the recycling of nitrogen to resynthesise amino acids in the cell. Additionally, the photorespiratory cycle, as occurs in leaves as a result of the oxygenase activity of RUBISCO, also generates ammonium ions, which can be re-assimilated here. Glutamine synthetase is encoded by a nuclear gene and is imported into the plastid, although a cytosolic form also exists. Subsequently, the GOGAT enzyme converts a molecule of glutamine and a molecule of 2-oxoglutarate to two molecules of glutamate (Fig. 7.4). Two different types of GOGAT enzyme are present in plastids. One type (Fd-GOGAT) uses ferredoxin as a source of reductant for the reaction and is used primarily in photosynthetic chloroplasts. Another type (NADH-GOGAT) uses NADH as a source of reductant and is primarily used in non-green plastids, such as those in root cells, or in young plastids which are developing and yet to become fully photosynthetic, such as in non-expanded non-green leaves. The end result of the GOGAT cycle is the synthesis of the amino acid glutamate, which can then be interconverted into the amino acids alanine and aspartate by the group of enzymes termed aminotransferases. Isoforms of aminotransferases are found in plastids specifically, as well as in the cytosol.

Four other amino acids: lysine, methionine, threonine and isoleucine, are synthesised by a common pathway, which utilises the amino acid aspartate as a starting point. The subsequent pathway, which synthesises these four amino acids, takes place almost entirely in the plastid except for the last step which synthesises methionine (Fig. 7.5). All of the enzymes are nuclear-encoded and contain transit peptide sequences enabling entry into the plastid. Many of the enzymes are sensitive to feedback control by the final amino acids. The first enzyme in the pathway, aspartate kinase, occurs in two different forms. One form is a single enzyme, which is sensitive to lysine and is down-regulated by high lysine concentration. The second form is a part of a bifunctional enzyme, which also has activity as a homoserine dehydrogenase. In this case, both of these enzyme activities on the same enzyme are sensitive to feedback control by high concentrations of threonine. The synthesis of lysine from dihydrodipicolinate (Fig. 7.5) involves six further enzyme steps, which are largely uncharacterised in plants. The synthesis of the sulphur

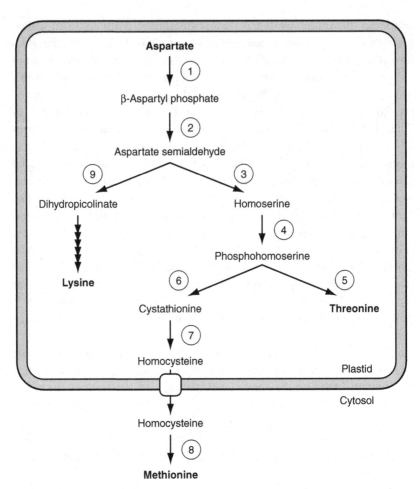

Fig. 7.5. The biosynthetic pathway for the amino acids methionine, lysine and threonine starts with the amino acid aspartate. All of this pathway operates inside the plastid, with the exception of the last step in methionine biosynthesis, where the conversion of homocysteine to methionine by methionine synthase occurs in the cytosol. Enzymes are (1) aspartate kinase, (2) aspartate semialdehyde dehydrogenase, (3) homoserine dehydrogenase, (4) homoserine kinase, (5) threonine synthase, (6) cystathione-γ-synthase, (7) cystathione-β-lyase, (8) methionine synthase, (9) dihydropicolinate synthase.

containing amino acid methionine occurs from the precursor homocysteine, which is exported from the plastid and converted to methionine by the cytosolic enzyme methionine synthase. Methionine is then subsequently imported into the plastid. Threonine can then form the starting

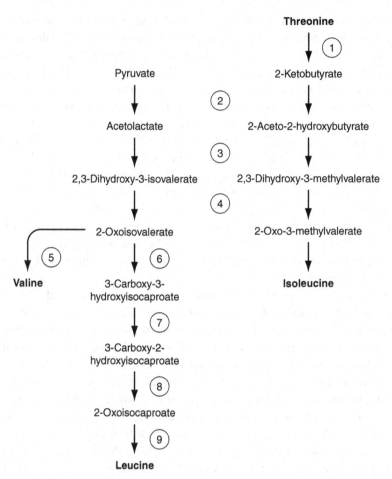

Fig. 7.6. The biosynthetic pathway of the branched amino acids valine, leucine and isoleucine share common enzymes in part of the pathway, isoleucine being synthesised from threonine and valine and leucine being synthesised from pyruvate. Enzymes are (1) threonine dehydratase, (2) acetohydroxyacid synthase, (3) ketoacid reductoisomerase, (4) dihydroxyacid dehydratase, (5) aminotransferase, (6) isopropylmalate synthase, (7) isopropylmalate dehydrogenase, (8) isopropylmalate dehydrogenase, (9) aminotransferase.

point for the synthesis of isoleucine, a pathway that has several enzymes in common with the biosynthetic pathway of valine and leucine (Fig. 7.6) from the starting point of pyruvate, an abundant metabolite in the plastid. Once more, this entire pathway occurs in the plastid with all of the enzymes nuclear-encoded and plastid-targeted.

The aromatic group of amino acids, containing the amino acids phenyl-alanine, tyrosine and tryptophan are synthesised in the plastid by a complex pathway of reactions, which includes the molecule shikimate and hence is termed the shikimate pathway (Fig. 7.7). The pathway commences with the condensation reaction joining erythrose-4-phosphate and phosphoenol pyruvate via the enzyme DHAP synthase (Fig. 7.7). Erythrose-4-phosphate is an intermediate in the photosynthetic Calvin cycle and in illuminated chloroplasts up to 30% of the carbon fixed by RUBISCO is thought to leave the Calvin cycle by this means and is used in the shikimate pathway. The key molecule, which precedes the synthesis of these three aromatic amino acids, is chorismate, which is synthesised from this condensation reaction in seven enzymatic steps (Fig. 7.7). From chorismate, a further four enzymatic steps give rise to tyrosine and phenylalanine, whereas tryptophan is synthesised by a separate branch incorporating six enzyme reactions, which gives rise to indol and subse-quently tryptophan.

Once more all of these enzymes are nuclear-encoded and plastid-targeted. One feature of the biosynthetic pathways for amino acids in plastids, which has been elucidated from the sequencing of plant nuclear genomes, is that different isoforms of some of these enzymes occur, encoded by different members of nuclear gene families and that these enzyme isoforms show different patterns of tissue-specific expression. Thus, plastids in different tissue types use subtly different versions of some of these amino acid biosynthetic enzymes. In addition, there are other levels of complexity in that feedback control by the final product on some of the biosynthetic steps is commonplace, thereby regulating the steady-state levels of amino acids produced and preventing their over-production. The enzyme EPSP synthase (enzyme 6 in Fig. 7.7) is the target for the widely used herbicide glyphosate, which effectively prevents aromatic amino acid biosynthesis in the plastid and thereby is ultimately lethal to the plant. EPSP synthase has been the target of much research and genetic engineering, including expression on the plastid genome (see Chapter 9). Not only does the shikimate pathway synthesise these three aromatic amino acids but it is also the mode of synthesis of a wide variety of secondary metabolites in the cell, arising from either further metabol-ism of these aromatic amino acids or via branch points in the pathway from intermediates.

The amino acid histidine is synthesised by a specific pathway involving eight enzymatic steps, commencing with the reaction between phospho-ribosyl pyrophosphate and ATP. The presence of transit peptide sequences

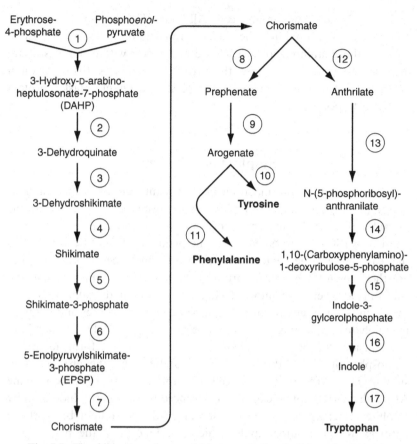

Fig. 7.7. The shikimate pathway, which occurs in the plastid, synthesises not only the three aromatic amino acids, tyrosine, phenylalanine and tryptophan, but is also the starting point for the biosynthesis of many different secondary metabolite molecules including aromatics, alkaloids and lignin. A key step is the synthesis of chorismate, from which the pathway branches to give rise to the different amino acids. Enzymes are (1) DAHP synthase, (2) dehydroquinate synthase, (3) dehydroquinate dehydratase, (4) shikimate dehydrogenase, (5) shikimate kinase, (6) EPSP synthase, (7) chorismate synthase, (8) chorismate mutase, (9) prephenate aminotransferase, (10) arogenate dehydrogenase, (11) arogenate dehydratase, (12) anthranilate synthase, (13) p-ribosyl anthranilate transferase, (14) phosphoribosyl anthranilate isomerase, (15) indoglycerolphosphate synthase, (16) trytophan synthase α, (17) tryptophan synthase β.

in these histidine synthetic enzymes implies that histidine is synthesised solely within the plastid compartment. Of those remaining amino acids required for protein synthesis, asparagine and proline are synthesised in the cytoplasm and serine and glycine are synthesised in the peroxisomes and mitochondria within the cell. All four of these amino acids are subsequently imported into the plastid for plastid protein synthesis. Likewise, all those amino acids synthesised exclusively within the plastid are exported to some extent into the cytosol to enable cytosolic protein synthesis, whilst some remain in the plastid to enable plastid protein synthesis to occur.

Sulphur metabolism

Sulphur is an important macronutrient in plants and its primary assimilation as sulphate ions into plant metabolism occurs mostly in the plastid, primarily by incorporation into the amino acid cysteine. Sulphate ions (SO_4^{2-}) are taken up by the plant's roots and transported to the leaves where they enter the cell via a sulphate permease (SP1) and are then taken up by the chloroplast via a sulphate transporter in the chloroplast envelope membrane (Fig. 7.8). Sulphur dioxide, present in the air in polluted environments, can be taken up by leaves and oxidised to sulphate and forms an alternative source for sulphur acquisition. Once inside the plastid, sulphate reacts with ATP to form adenosine 5-phosphosulphate (APS), a reaction catalysed by the enzyme ATP sulphurylase. APS is then reduced to sulphite (SO_3^{2-}) by the enzyme APS reductase. Sulphite can then be reduced further to sulphide (S^{2-}) by sulphite reductase (Fig. 7.8) or incorporated into a lipid biosynthetic pathway to yield sulpholipids, which are unique to the plastid and play an important role in both the envelope membrane and the thylakoid membrane. This lipid biosynthetic pathway combines sulphite with UDP-glucose to form UDP-sulphoquinovose, a reaction driven by the enzyme SQD1 synthase (Fig. 7.8). UDP-sulphoquinovose is then coupled with diacylglycerol to form the lipid sulpholipid SQDG (sulpholipid 6-sulpho-α-D-quinovosyl diacylglycerol). The extent of sulpholipid biosynthesis is significant, with up to one-third of all the sulphur contained in a leaf incorporated in these plastid-specific lipids. If not incorporated into sulpholipids, sulphide can be incorporated into the amino acid cysteine by combining with O-acetylserine (OAS), which is generated from the amino acid serine by serine acetyltransferase (SAT) (Fig. 7.8). The incorporation of sulphide with OAS is mediated by the enzyme OAS thiol lyase. The production of the amino acid cysteine represents the major end point of sulphur assimilation, and from where cysteine can be incorporated into either plastid or cytosolic proteins or can be

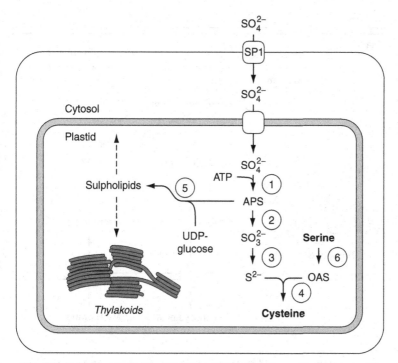

Fig. 7.8. Sulphur metabolism in plant cells takes place mostly in the plastid via the assimilation of sulphate ions to generate the amino acid sulphur-containing amino acid cysteine. The import of sulphate ions across the plastid envelope membrane is assumed but the transporter is uncharacterised. In this pathway, the reaction of sulphite ions with UDP-glucose leads to the production of sulpholipids in the plastid, which are important in the function of the plastid envelope and thylakoid membranes. APS = adenosine 5-phosphosulphate, OAS = O-acetylserine, SP1 = sulphate permease. Enzymes are (1) ATP sulphurylase, (2) APS reductase, (3) sulphite reductase, (4) OAS thiol lyase, (5) SQD1 synthase, (6) serine acetyl transferase.

converted to the other sulphur-containing amino acid, methionine (Fig. 7.9) via the intermediate homocysteine.

The role of cysteine in proteins is crucial since cysteine residues form disulphide bonds and play a major role in establishing protein three-dimensional structure. In the plastid, disulphide bonds are important in many enzymes which are light activated by thioredoxin, which itself contains a disulphide bond (see Chapter 4). In addition to being incorporated into proteins, the amino acids, cysteine and methionine, are both

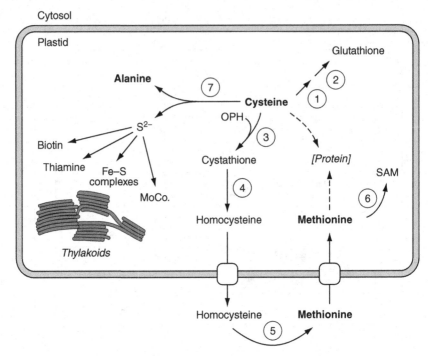

Fig. 7.9. The amino acid cysteine is central to sulphur metabolism in the plastid, giving rise to the other sulphur-containing amino acid, methionine, which is synthesised in the cytosol and imported for plastid protein synthesis. Cysteine is also the basis for synthesising glutathione, which functions in stress responses in the cell. Cysteine can also generate sulphide ions (S^{2-}), which can form part of iron–sulphur complexes and several other co-enzymes in the plastid. OPH = O-phosphohomoserine, SAM = S-adenosyl methionine, MoCo = molybdenum co-factor. Enzymes are (1) glutathione synthetase, (2) γ-glutamylcysteine synthetase, (3) cystathione-γ-synthase, (4) cystathione-β-lyase, (5) methionine synthase, (6) S-adenosyl methioine synthetase, (7) cysteine desulphurase.

starting points for the synthesis, in the plastid, of other sulphur compounds that are either used within the plastid or exported to the cytosol. Cysteine is crucial to the synthesis of glutathione, a tripeptide of glutamate–cysteine–glutamate (γ-Glu-Cys-Glu). Firstly, Glu and Cys are joined by the enzyme γ-glutamylcysteine synthetase, and subsequently another Glu is added by the enzyme glutathione synthetase (Fig. 7.9). This synthesis occurs in the plastid, as well as in the cytoplasm. Two glutathione molecules can be oxidised to form one molecule of oxidised glutathione, a process that yields electrons and protons for the reduction of other cellular components. Reduced glutathione can be regenerated by the enzyme glutathione reductase, which uses NADPH as its source of protons and electrons. Glutathione plays an important role in the plastid as an antioxidant and in its response to stress, especially in processing hydrogen peroxide, generated by the production of oxygen radicals in photosynthesis and in the binding of heavy metal ions. Cysteine can be converted into methionine by three enzymatic steps, the last of which appears to occur in the cytosol (see previous section). Methionine is then reimported into the plastid (Fig. 7.9). Much of the methionine synthesised in the cytosol is converted to S-adenosyl methionine (SAM), which is then utilised in the cytosol or in the plastid after being imported, as a methyl donor in a variety of metabolic reactions, which give rise to polyamines and the important gaseous plant hormone, ethylene.

Cysteine itself can be converted to alanine by the enzyme cysteine desulphurase (Fig. 7.9), a reaction that generates reduced S for the production of several different co-enzymes found in the plastid as well as the reduced S, which is incorporated into the Fe–S clusters found in the photosynthetic electron transport protein complexes of the thylakoid membrane (see Chapter 4).

Fatty acid biosynthesis

Lipids are of major importance in plant cells, forming the core components of membranes as well as providing storage materials within cells that store lipids for energy, such as in lipid-storing seeds. A major component of lipid molecular structures are fatty acids, which in plant cells are synthesised entirely within the plastid compartment. The fatty acid biosynthetic mechanism in plastids is prokaryotic in nature and similar to that found in modern-day bacteria, which is capable of producing acyl chains up to 18C in length. The key molecule from which fatty acids are made is acetyl CoA, which is carboxylated by the enzyme acetyl CoA carboxylase to form malonyl CoA. Acetyl CoA is unable to be imported across the plastid envelope so it needs to be synthesised from other metabolites, which can be imported. It would appear that various different intermediate metabolites can act as a carbon source for synthesising acetyl CoA in heterotrophic plastids, in which the majority of storage fatty acids are made, such as those in seed endosperm or cotyledons, where fatty acids are a major storage product. Pyruvate, triose phosphate, glucose 6-phosphate, malate and other metabolites can all be used as a

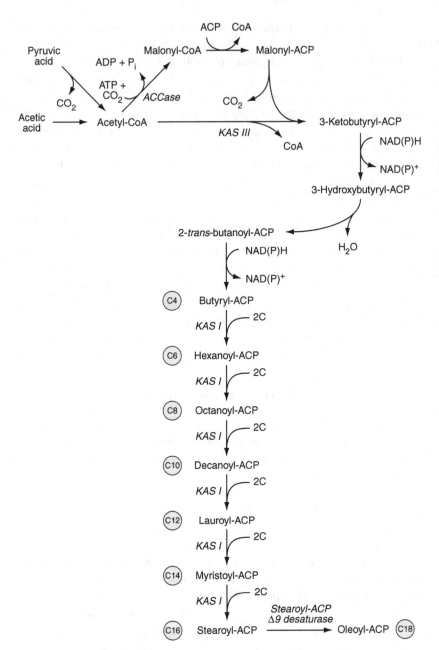

Fig. 7.10. The pathway of fatty acid biosynthesis that takes place in plastids. The starting molecule is acetyl-CoA, which is converted to malonyl CoA by the enzyme acetyl CoA carboxylase. Malonyl CoA eventually yields the two carbon atoms that elongate the acyl chain by 2C in each step, and the

source for biosynthesis of acetyl CoA in the plastid. Acetyl CoA carboxylase is a complex enzyme in most dicotyledonous plants, made of four different subunits encoded by different genes. These subunits are biotin carboxylase, biotin carboxyl carrier protein and two subunits of carboxyltransferase, one of which is encoded in the plastid genome by the *accD* gene (see Chapter 3). A second version of the acetyl CoA carboxylase enzyme also exists as a homomeric form encoded by a nuclear single gene, the product from which is plastid-targeted. Fatty acids are synthesised by a multi-enzyme complex, called fatty acid synthetase (FAS), which resides in the plastid stroma and contains a small acyl carrier protein (ACP) which facilitates the elongation of the growing acyl chain. It combines with malonyl CoA to form malonyl-ACP, catalysed by the enzyme malonyl-CoA:ACP malonyl transferase, and malonyl-ACP then acts to elongate the acyl chain (Fig. 7.10). Elongation of the chain is catalysed at different stages by β-ketoacyl-ACP synthase enzymes, termed KAS. Thus KASI, KASII and KASIII catalyse different parts of the elongation pathway (Fig. 7.10). The stroma of the plastid contains a desaturase enzyme which introduces a C=C at the C9 position in the stearoyl-ACP molecule, giving rise to oleoyl-ACP (Fig. 7.10). This is the most abundant fatty acid made by the plastid fatty acid biosynthetic pathway. Other membrane-bound desaturases within the plastid can give rise to a spectrum of unsaturated fatty acids, with C=C bonds at differing positions in the acyl chain.

The incorporation of these different fatty acids into glycerolipids can occur either within the plastid compartment, using a basically prokaryotic synthetic mechanism or in the endoplasmic reticulum via a eukaryotic-type mechanism, the latter case requiring the export of fatty acids from the plastid and import into the endoplasmic reticulum. Thus, the array of chloroplast lipids, which are crucial to envelope membrane and thylakoid membrane function can be synthesised (see Chapter 6). The exact nature

Caption for Fig. 7.10. (cont.)
reactions leading to the elongation step require binding with the acyl-carrier protein (ACP). The first condensation reaction of acetyl CoA and malonyl-ACP, giving rise to a 4C molecule, is catalysed by KASIII. 2C from this 4C molecule are eventually transferred to the growing acyl chain in each step. Different reactions in the elongation process are catalysed by KASI and KASII enzymes. KAS = β-ketoacyl-ACP synthase. The final outcome is fatty acids containing 16C or 18C acyl chains. Desaturase enzymes can introduce double C=C bonds in the chain. The enzyme stearoyl-ACP Δ9-desaturase introduces a C=C at position 9 producing oleoyl-ACP.

of how fatty acids are trafficked to the endoplasmic reticulum remains unclear, although the extent of fatty acid export from the plastid is very significant. In leaf mesophyll cells and tissues containing non-green plastids, in the order of 90% of the fatty acids made in the plastid are exported to the endoplasmic reticulum. However, one mechanism by which fatty acids can be exported from the plastid is by their activation to an acyl-CoA form by enzymes called acyl-coenzyme A synthetases. A plastid-targeted form of acyl-coenzyme A synthetase resides on the plastid envelope and facilitates fatty acid acyl-CoA production and subsequent export. The exact nature of how lipids subsequently synthesised in the endoplasmic reticulum return to the plastid remains to be discovered. The movement of lipids synthesised on the plastid envelope and destined for the thylakoid membrane are likely moved via vesicle trafficking as described in Chapter 6.

Carotenoid biosynthetic pathway

Carotenoids are a large group of molecules, which are synthesised largely in plastids in higher plants and have several roles in photosynthetic light harvesting on the thylakoid membrane and photoprotection as well as in providing coloured pigments in plant tissues in chromoplasts (see Chapter 2). They can be observed in coloured ripe fruit, such as tomatoes and peppers as well as being visible in the autumn during leaf senescence, as they are revealed after the chlorophyll, which normally masks them, is degraded. The carotenoid biosynthetic pathway also synthesises molecules that are used in a variety of biosynthetic pathways leading to various groups of complex molecules used within the plastid and also in the cytosol. The pathway commences with the synthesis of two isoprene isomers, isopentenyl diphosphate (IPP) and dimethylallyl diphosphate (DMAPP) (Fig. 7.11). These molecules also provide the basis for other biosynthetic pathways making important biological molecules in plastids including monoterpenes, plastoquinones, tocopherols and phylloquinones. Consideration of all of these complex biosynthetic pathways is beyond the scope of this book, but information about these pathways is mentioned in the further resources at the end of this book.

The carotenoid biosynthetic pathway requires several different enzymes, all of which are nuclear-encoded and imported into the plastid. IPP and DMAPP are made by combining glyceraldehyde 3-phosphate and pyruvate to form deoxy-D-xylulose 5-phosphate (DXP), catalysed by

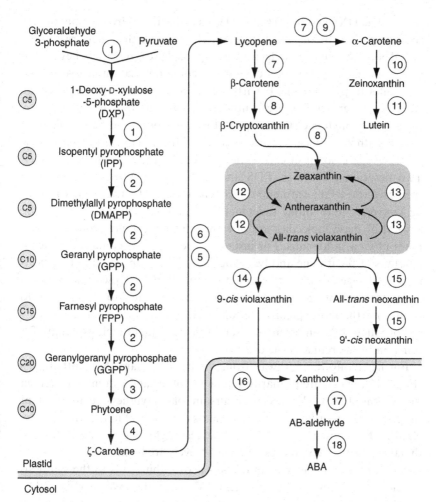

Fig. 7.11. The pathway of carotenoid biosynthesis that takes place mostly in the plastid. In addition to synthesising various carotenoids used in photosynthetic complexes and in chromoplasts as pigments, the pathway eventually synthesises the plant hormone abscisic acid (ABA) in the cytosol. Enzymes are (1) deoxy-D-xylulose 5-phosphate synthase, (2) isopentyl isomerase, (3) phytoene synthase, (4) phytoene desaturase, (5) ε-carotenoid desaturase, (6) carotenoid isomerase, (7) lycopene β-cyclase, (8) β-carotene hydroxylase, (9) lycopene ε-cyclase, (10) β hydroxylase, (11) ε-hydroxylase, (12) zeaxanthin epoxidase, (13) violaxanthin de-epoxidase, (14) violaxanthin isomerase, (15) neoxanthin synthase, (16) 9-*cis* epoxycarotenoid dioxygenase, (17) short chain dehydrogenase, (18) AO aldehyde oxidase. (Redrawn from figure, courtesy of Dr. Ganesh Balasubramanian.)

the enzyme DXP synthase (Fig. 7.11). The addition of five carbon units in sequential steps by the enzyme isopentyl isomerase then gives rise to the 20-carbon molecule geranylgeranylpyrophosphate (GGPP). A key step in this biosynthetic pathway is the subsequent condensation of two GGPP molecules to produce the 40-carbon molecule phytoene, from which all subsequent carotenoids are synthesised. This reaction is catalysed by the enzyme, phytoene synthase, and this step appears to be rate limiting in the pathway, since over-expression of the phytoene synthase gene increases the amount of carotenoids formed at the ends of the pathway. Two desaturase enzymes, PDS and ZDS, and a carotenoid isomerase then give rise to the carotenoid lycopene (Fig. 7.11). Lycopene can then be converted into other carotenoids including β-carotene and lutein. Several of these molecules are of significant importance from a dietary point of view in humans. For instance, ripe tomatoes contain significant amounts of β-carotene and lycopene, which are orange and red, respectively, and together gives them their colour. Both of these molecules can act as antioxidants in the cell and have been postulated to be beneficial to health. Furthermore, β-carotene can be cleaved to form retinal, a precursor of vitamin A, an essential vitamin in the human diet, especially for function of the retina in the eye.

Further along the pathway from the carotenes are the xanthophylls (Fig. 7.11), which are primarily yellow in colour and can interconvert in the xanthophyll cycle, which occurs in photosynthetic complexes of the thylakoid in order to process toxic oxygen radical ions formed there. Xanthophylls are also involved in excess light energy dissipation in thylakoid protein complexes when light levels are high.

The export from the plastid of 9-*cis* neoxanthin leads to the synthesis of the plant hormone abscisic acid (ABA) in the cytosol (Fig. 7.11), which plays a major role in drought responses and seed germination. The regulation of carotenoid biosynthesis is undoubtedly complex, but in general appears to be controlled by the extent of gene expression of many of the enzymes involved. The pathway is tightly regulated in response to environmental stimuli in chloroplasts since the relative proportions of different carotenoids present on the thylakoid membrane will have a profound effect on the chloroplast's response to a changing photoenvironment.

8

Plastids and cellular function

From what has been discussed in the preceding chapters, it should be clear that the plastid is a highly dynamic organelle in terms of its biochemistry, its molecular biological systems, its photochemistry and its molecular interaction with the nucleus. In addition to these aspects of plastid biology, one also needs to consider how the plastid resides within the cytoplasm of the cell, in many cases as a large closely packed population of individual organelles resulting from extensive divisions, and how the individual organelles move about in the cell and physically interact with other cellular organelles. Classic images of plastids from sectioned leaves and electron micrographs (see Figs. 2.4, 2.5) give the impression of a static organelle in which little changes physically with time. Modern imaging techniques, together with developments in the use of visible molecular markers that can be artificially incorporated into plastids, reveal a very different view of plastids: that of a highly dynamic organelle capable of rapid changes in morphology and movement within the cell. In this chapter, some of these dynamic aspects of plastid biology are described. In addition, we consider how different types of plastids differentiate in different types of cells and how they might influence plant development itself.

Plastid division

All plastids in a plant originate from proplastids in the cells of the meristem, which in turn are generally derived from the few proplastids in the egg cell in the flower. The fact that plastids need to divide to establish large populations of plastids in the large number of cells that make up a plant is obvious. The understanding of plastid division started with the observation of plastids in meristem cells and in young developing

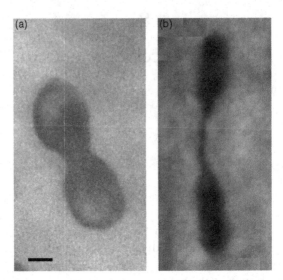

Fig. 8.1. Chloroplasts from seedlings of *Arabidopsis* observed during their division process. (a) A young dividing chloroplast in an expanding leaf mesophyll cell showing the central constriction and the characteristic dumbbell shape, typical of dividing plastids. (b) A plastid from an epidermal cell in the latter stages of division, showing a long thin isthmus joining the two plastid bodies together. The isthmus will eventually break, the membranes will fuse and two separate daughter plastids will be formed. Scale bar = 1 μm.

leaf mesophyll cells showing central constrictions making them dumbbell shaped (Fig. 8.1). These were considered to be in the early stages of a division process, and careful analysis of these plastids showed that they eventually pinch off in the middle of the plastid to form two daughter plastids by a process termed binary fission. This process of plastid division happens in meristem cells where proplastids divide to ensure continuity in cells after cell division (see Chapter 2). Plastid division also occurs in many cell types where a significant population of plastids is required in the mature cell. The best characterised pathway of plastid division is that of young chloroplasts that develop in mesophyll cells of leaves and which divide as the cell expands to generate significant populations of chloroplasts within the cell. The mode of their division appears to be similar to that seen in proplastids with a central constriction, which can form into a long thin isthmus joining the two plastids in the later stages of division, which is then followed by separation of two daughter chloroplasts (Fig. 8.1). Newly replicated chloroplasts then need to increase in size by growth before a further round of division can occur. In an average leaf mesophyll cell, there are about 20 or so proplastids at the beginning of the cell's development and, at maturity, leaf mesophyll cells contain between 100 and 150 chloroplasts. Thus, between three and four rounds of division occur during the accumulation of a chloroplast population in a leaf mesophyll cell. There is, however, much variation between mesophyll cells in the actual number of chloroplasts that they contain and normally there exists a good correlation between the size of the mesophyll cell and the number of chloroplasts within it. In such mesophyll cells, the chloroplasts reside in the thin sheet of cytoplasm, which is pressed against the cell wall by turgor pressure from the vacuole. Thus, the chloroplasts form a monolayer of adjacent chloroplasts, which cover the inside of the cell wall, within the cytoplasm bounded by the plasma membrane on the outside and by the tonoplast membrane on the inside. In mesophyll cells that develop in well-lit conditions, the chloroplast population covers around 70% of the cell's surface area (see Fig. 2.3).

Early observation of electron micrographs of dividing proplastids and chloroplasts showed an electron-dense ring, termed the plastid dividing ring (PD), forming around the centrally constricted membranes with parts on both the outside and the inside of the plastid envelope membrane (Fig. 8.2). Similarities in the way in which proplastids and chloroplasts were observed to divide led to the assumption that the two mechanisms are basically similar. Significant progress in understanding the mechanism of plastid division has come about by two routes. Firstly, genes were found in higher plant nuclear genomes which are homologues of genes originally involved in bacterial cell division, but which now reside in the plant nucleus and are involved in plastid division. Secondly, other nuclear genes were identified by characterising mutant plants, in which the extent of plastid division is altered and thereby plastid populations in mesophyll cells show differences in size. Both of these approaches have been fruitful and several proteins have been identified, which function in various parts of what is obviously a complex plastid division mechanism. A major protein in the plastid division mechanism is the protein filamentous temperature-sensitive Z (FtsZ). FtsZ functions in bacterial cell division and two homologues of the bacterial gene are present in plant nuclear genomes, with both FtsZ proteins residing inside the plastid (Fig. 8.3). FtsZ proteins have characteristics of the cytoskeletal protein tubulin and they can form filaments, which form a contractile ring on the inside of the

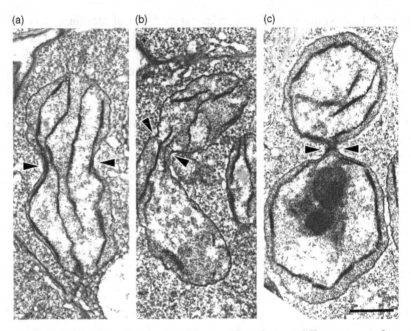

Fig. 8.2. Electron micrographs of young chloroplasts at different stages of their division process in expanding mesophyll cells in seedlings of *Arabidopsis*. (a) The beginning of plastid constriction at a central location along the long axis of the plastid. (b) Further constriction causes a narrowing of the plastid at its midpoint. (c) The last stage of plastid division with a narrow isthmus joining the two daughter plastids. A dark area is visible at the isthmus, which is the plastid dividing ring, which only becomes visible at the latter stages of the division process. Scale bar = 1 μm. (From Robertson EJ, Rutherford SM, Leech RM (1996). *Plant Physiology* 112, 149–159. © American Society of Plant Biologists and reproduced with permission.)

plastid envelope and, during the progression of the division process, the ring contracts, pulling the plastid envelope membrane inwards (Fig. 8.3). Eventually, the constriction of the envelope membrane results in fusion of the inner and outer envelope membranes separately and two separate daughter plastids are formed from the original plastid. In many examples of plastid division, the central constriction can take the form of a long thin isthmus, which continues to join the two daughter plastids together until eventual separation.

The FtsZ ring is stabilised by the protein ARC6, which was first identified by characterisation of a mutant of *Arabidopsis*. The leaf mesophyll cells in the *arc6* mutant have a spectacular phenotype in that they contain only one or two huge giant chloroplasts instead of the large population of normal-sized chloroplasts. The generation of force by the constricting FtsZ ring is facilitated by its interaction with the protein ARC5, also originally characterised from an *Arabidopsis* mutant. ARC5 has similarities to the protein dynamin, which functions in several different cell biology processes as a motor protein. In beautifully elegant experiments, the FtsZ ring can be isolated and ARC5 protein added, which induces ring constriction *in vitro* and the force generated can be measured using laser-based optical tweezers. Two further proteins, PDV1 and PDV2, link the outer plastid division ring to the outer envelope membrane. Several other proteins are also implicated in the division process, although the exact manner in which they come together to carry out the constriction is somewhat unclear, especially how the FtsZ ring relates to the PD rings (Fig. 8.4). Although the PD ring is the only visible component of the division process, as seen in the electron microscope (Fig. 8.2), it forms a short while after the FtsZ ring has been formed and locates correctly at the midpoint of the plastid. It is only after the PD rings have formed that visual constriction of the plastid commences. Although PD rings can be isolated intact, their exact constituents are unknown, although they do have a filamentous structure. In some instances, a third PD ring forms within the lumen of the envelope membrane.

An important aspect of the plastid division process is that it occurs at the midpoint of the long plastid axis and thereby gives rise to two equally sized daughter plastids. The fact that a mechanism exists to ensure that this central constriction is located correctly shows that plastids should not be considered simply as membranous sacs but that they are clearly polarised organelles with long and short axes and distinct poles at either end. The molecular machinery, which ensures a centrally placed constriction, is based on the system of Min proteins, which functions in bacteria to ensure a centrally placed division event in the bacterial cell. The proteins MinD and MinE both function to ensure that the FtsZ ring forms at the midpoint of the long axis and not at either of the poles. MinD and MinE form a complex, which is stimulated by Ca^{2+} ions to induce ATPase activity by the MinD protein. Exactly how this complex prevents the FtsZ ring forming in the wrong place is unclear, although another component of the complex, ARC3, can interact with FtsZ proteins and may provide a connection between the Min complex and the FtsZ ring.

There is still much to be revealed about the full spectrum of components in this plastid division mechanism and how they all work together,

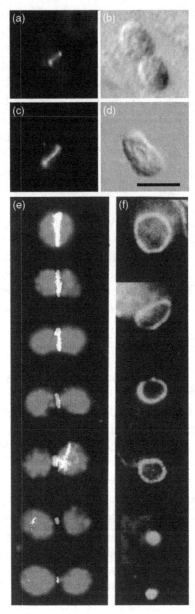

Fig. 8.3. Rings formed by the FtsZ proteins during chloroplast division. In
(a)–(d), the two different FtsZ proteins associated with plastid division have
been immunolocated in isolated dividing chloroplasts and both are shown to
be located around the midpoint of the dividing chloroplast in the same place
as the central constriction. The isolated chloroplast is shown in (b) and

but it is interesting that the division mechanism that functions in modern-day chloroplasts constitutes a mixture of those proteins derived evolutionarily from division proteins functional in bacteria and newly evolved proteins, which are only functional in plant plastid division and not in bacteria. Although the binary fission mechanism is regarded as the main way by which plastids divide, different strategies for division may also exist. A budding mechanism, by which small plastid bodies bud off from the main organelle, has been observed occasionally and may constitute an alternative mechanism in some plastid types.

Although the way in which chloroplasts divide is only partially understood, other types of plastid are capable of division to a limited extent. For instance, during the development of cells in potato tubers and the laying down of starch in the amyloplasts in those cells, amyloplasts divide. In a contrasting system, during the proliferation of cells in plant tissue-cultured systems, leading to masses of callus tissue, unpigmented leucoplasts divide. However, in other systems, such as in chromoplast differentiation from chloroplasts in fruit ripening, the generation of large populations of plastids occurs at the chloroplast stage by chloroplast division, each chloroplast then differentiating into a chromoplast.

Dynamics of plastid morphology

The majority of mature chloroplasts in leaf mesophyll cells appear to be a relatively consistent shape when viewed by the light microscope or by the electron microscope as discussed in Chapter 2. However, chloroplasts and other types of plastids are capable of taking up different morphologies

Caption for Fig. 8.3. (cont.)
(d) and the corresponding fluorescencent image is shown in (a) and (c) for FtsZ1 (a) and FtsZ2 (c). (e) shows the FtsZ ring associated with chloroplasts at different stages in the division process from the earliest stage at the top to the final stage at the bottom. (f) Shows rings of FtsZ isolated from similar stages of the chloroplast division process to that shown in (e). Note how the FtsZ ring constricts and gets smaller as the division process proceeds. Scale bar = 5 μm. ((a)–(d) reprinted with permission from the *Annual Review of Plant Physiology and Plant Molecular Biology* 52, 315–333 by Annual Reviews © Annual Reviews 2001 http://www.annualreviews.org. (e)–(f) reprinted from Kuroiwa H, Mori T, Takahara M, Miyagishima S, Kuroiwa T (2002). Chloroplast division machinery as revealed by immunofluorescence and electron microscopy. *Planta* 215, 185–190 with kind permission from Springer Science + Business Media.)

(a)

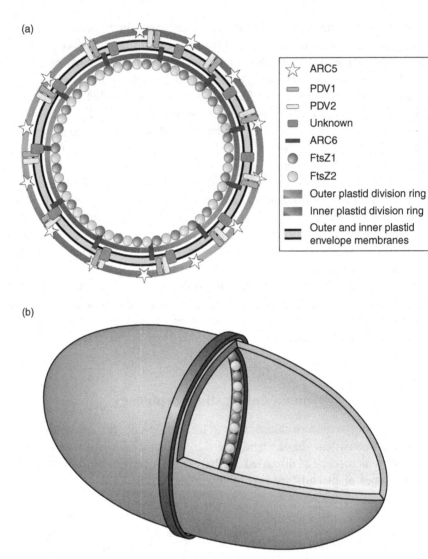

☆	ARC5
⬭	PDV1
⬭	PDV2
⬛	Unknown
▬	ARC6
●	FtsZ1
○	FtsZ2
▬	Outer plastid division ring
▬	Inner plastid division ring
═	Outer and inner plastid envelope membranes

(b)

Fig. 8.4. A model of how the plastid division machinery is arranged at the midpoint of the dividing plastid and is composed of concentric rings of an FtsZ ring and plastid dividing rings, interlinked with proteins and associated with the inner and outer plastid envelope membranes. (a) Shows how the various components of the mechanism are associated with the plastid envelope membrane and (b) shows how distinct rings are present on both the inner and outer surface of the plastid envelope.

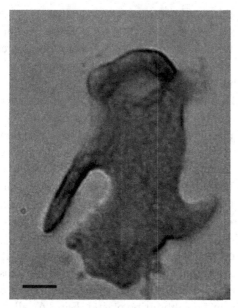

Fig. 8.5. An isolated chloroplast of the *arc6* mutant of *Arabidopsis*, which maintains its irregular three-dimensional shape. These *arc6* chloroplasts are significantly bigger than normal wild-type chloroplasts. Scale bar = 10 μm.

and it has become clear that the plastid should be regarded as a highly dynamic organelle. A good example of how chloroplasts can become highly irregular in morphology is seen in the *arc6* mutant of *Arabidopsis*, in which plastid division is perturbed and the resulting few chloroplasts in leaf mesophyll cells become greatly enlarged and take up a complex three-dimensional morphology, with extended protuberances of the plastid body and an irregular outline (Fig. 8.5). If isolated from the cell and suspended in aqueous buffer, these *arc6* chloroplasts retain their three-dimensional shape and do not revert to a spherical or oblate spheroid shape as might be expected. This strongly suggests that some mechanism internal to the plastid envelope membrane exerts control over the three-dimensional shape of the plastid. Exactly what this mechanism might be or how it functions is unclear. The idea of a plastoskeleton, analogous to the cytoskeleton which functions in the cytoplasm, has been suggested, but precise components of its make-up, if it exists, are unknown.

Extreme morphological changes in plastid body can also be seen clearly when the recombinant protein green fluorescent protein (GFP) (see Chapter 9) is targeted to the plastid in transgenic plants. When such plastids are viewed by the fluorescence or confocal microscope and the GFP is

fluoresced, long thin protuberances, which emanate from the plastid body
into the cytoplasm, can be seen (Fig. 8.6). These membranous tubules,
which contain stroma, but no thylakoid membrane or chlorophyll are
called stromules (*stroma*-filled tub*ules*) and were originally seen by light
microscopy in the 1960s but never properly described as such, and were
subsequently rediscovered in the late 1990s. Stromules are mostly between
350 and 850 nm in diameter and are formed by outgrowths of both the
inner and outer plastid envelope membranes together. Some, if not all of
this extension, is powered by attachment of the stromule envelope membranes to the actin microfilament cytoskeleton network within the cytoplasm and the generation of force by myosin motor proteins associated
with the microfilament network. Binding of stromule membranes, either
at their tip or along their length, in this way enables them to be pulled out
and extended. Movement of stromules is surprisingly dynamic, in that
they can undergo extensive contortions in the cytoplasm including extension and retraction, branching and rejoining to form closed loop structures, breaking off into distinct vesicles and at the most extreme, fusing
with stromules from a neighbouring plastid (Fig. 8.6), all on a timescale of
seconds. Careful analysis of stromules emanating from plastid bodies,
either with the aid of GFP or with careful observation using the light
microscope, has revealed this significant aspect of plastid morphological
dynamics, which was previously unknown and consequently one should
change the way in which one considers plastids as a cellular organelle.
Rather than viewing the plastid as a defined static structure within the
cytoplasm, the plastid should be considered as a dynamic entity, forming
into complex morphological contortions with much flow of the plastid
envelope membranes to generate these stromule structures, and occasionally linking together individual plastids.

There is significant variation in the extent of stromule formation on
different types of plastids in different types of cells. Stromules appear
much more extensively on non-green plastids lacking chlorophyll, rather
than on mature green chloroplasts such as those in leaf mesophyll cells.
Thus on mesophyll cell chloroplasts, stromules are relatively rare,
whereas in cells containing poorly developed chloroplasts or leucoplasts,
such as in root cells, leaf hair cells or in tissues containing other plastid
types such as chromoplasts, stromules are relatively abundant (Fig. 8.6).
Indeed, in cultured tobacco cells, which contain leucoplasts, stromules
can form extensive arrays of tubules, which dwarf the main plastid body
and emanate from the plastid body toward the plasma membrane in a
network-like array (Fig. 8.6). In pericarp cells in ripening tomato fruit,

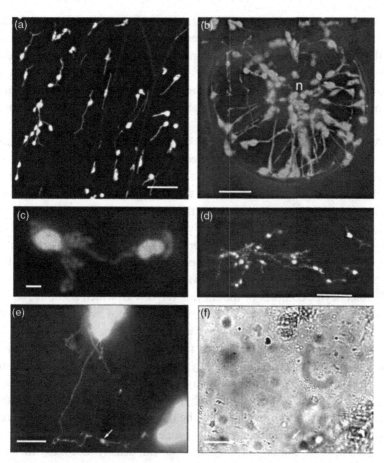

Fig. 8.6. Stromules observed in different types of cells and visualised using fluorescent protein (GFP) targeted to the plastid. (a) Plastids in the hypocotyl cells of tobacco seedlings show extensive stromules, often with two stromules arising from the same plastid at either pole. Scale bar = 10 μm. (b) Plastids and stromules in a cultured tobacco cell showing plastids arranged around the nucleus (n) linking via stromules to plastids at the cell periphery. In such cells plastids appear as a dynamic networked community rather than individual discrete organelles. Scale bar = 10 μm. (c) Two plastids in a hair cell from a tomato leaf joined together by a stromule consisting of many vesicle-like structures. Scale bar = 1 μm. (d) A complex array of plastids and networking stromules in a root cell of an *Arabidopsis* seedling. Scale bar = 10μm. (e),(f) Stromule linking two chromoplasts in a pericarp cell in a ripe tomato fruit. The brightfield picture in (f) shows the two chromoplasts which are joined by a long stromule with many vesicle-like structures associated, one of which is arrowed. Scale bar = 10 μm. ((b) reprinted from Kwok EY, Hanson MR (2004). Plastids and stromules interact with the nucleus and cell membrane in vascular plants. *Plant Cell Reports* 23, 188 with kind permission from Springer Science + Business Media.)

stromules associated with differentiating chromoplasts are long and extensive and highly beaded with vesicle-like structures, and these stromules appear able to produce separated vesicles, which themselves can differentiate into chromoplasts (Fig. 8.6). In some plastid types, such as root plastids, the plastid body is elongated and complex to start with and stromules on these plastids become an extreme part of the plastid's morphology (Fig. 8.6).

Although it is clear that stromules do not contain thylakoid membrane or chlorophyll, exactly which molecules or structures are capable of moving through a stromule is unclear. Certainly a small protein such as GFP, which is 30 kD in size, can move through stromules, and laser photobleaching experiments show that, when two plastids are joined together by a stromule, GFP can move from one plastid body to another. Similar conclusions come from experiments where GFP is microinjected into a plastid and subsequently appears in neighbouring plastids. The largest structure that has been shown to move through stromules is the full holoenzyme complex of the enzyme RUBSCO, which is about 500 kD in size but there is no evidence currently for the movement of ribosomes, DNA or RNA molecules into stromules and between plastids.

The exact role of stromules within the context of plastid function within the cell is unclear. They undoubtedly increase the surface area to volume ratio of plastids significantly and maybe are involved in enhancing transport across the plastid envelope. In hypocotyl cells of tobacco, the extent of stromule formation is related to the density of plastids within the cell such that, at high plastid density, stromules are less prevalent than when the cells are expanding rapidly and plastid density declines. In such situations, stromules become more extensive on plastids and individual stromules become longer. This suggests that stromules may be involved in a density sensing mechanism within the cytoplasm, monitoring the extent of the plastid population and moderating a plastid response.

A significant aspect of stromule biology is their ability to enhance interaction between plastids and other cellular organelles. This is clearly shown in the interaction between stromules and the nucleus, where in many types of cell with a relatively low density of non-green plastids, the plastids position themselves around the nucleus, in a peri-nuclear array (Fig. 8.6). Such an array is also seen in proplastids in dividing cells, an arrangement that ensures segregation of proplastids into both daughter cells at cytokinesis (see Chapter 2). In such arrays, stromules emanating from these plastids are shown to lie in grooves in the nuclear membrane, enabling remarkably close contact between the two organelles.

Consequently, when one considers plastid-nuclear signalling, in some situations, there is little cytoplasm separating the two organelles and their membranes are closely appressed, minimising the distance between the two organelles. In other situations, close physical interactions between stromules and other cellular organelles such as mitochondria or peroxisomes can be observed, although whether close physical presence such as this is a result of organelles binding to the same cytoskeletal tracks or whether these organelles can actually physically interact with each other in a functional way is unclear. The main point here though is that plastids are highly dynamic structures in terms of their morphology and via their stromules are capable of physical interaction not only with each other but also with other cellular organelles.

The plastid envelope membrane has the capacity to monitor its degree of stretch when it alters its morphology since it contains two mechanosensing proteins which can monitor membrane tension. These proteins, MSL2 and MSL3, are closely related to similar proteins in the bacterial plasma membrane, which monitor osmotic shock conditions. MSL proteins play a role in plastid division and, although it is not clearly understood, may monitor envelope membrane tension during constriction.

Plastid movement and positioning

In addition to possessing a dynamic morphology, plastids can also move around in the cytoplasm of the cell in a directed and controlled manner, largely in relation to the photoenvironment that the leaf or tissue experiences. Optimising plastid positioning in cells is one way in which the plant responds to changes in light intensity or direction, in order to optimise photosynthetic output. It has been known for over a century that chloroplasts in leaf cells can move and they show two different types of movement response according to the intensity of light. In low to moderate light conditions, chloroplasts in leaf mesophyll cells move to the cell walls most perpendicular to the direction of light so as to optimise their potential for light absorbance (Fig. 8.7). This movement is called the chloroplast accumulation response. In contrast, when the light intensity is high, the potential of the chloroplast for absorbing excess light energy and causing damage to the photosystems on the thylakoid membrane is greatly increased and so chloroplasts move to the side walls of the cell such that their profile area is orientated parallel to the direction of light and thus they minimise light interception. This movement is called the chloroplast

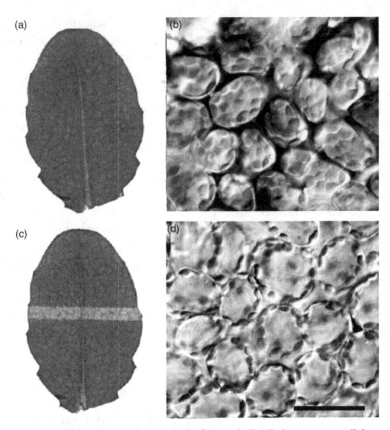

Fig. 8.7. Chloroplast movement in leaf mesophyll cells in response to light. When leaves are exposed to medium to low levels of light, the chloroplasts locate to the upper and lower surfaces of the palisade mesophyll cells forming layers perpendicular to the incident light falling on the leaf and thereby maximising their interception as in (b). When the leaf is covered and only a slit across the middle of the leaf is illuminated with high light, the chloroplasts move to the side walls of the cells, thereby minimising their interception and allowing more light to pass through the leaf as in (d). This results in a strip of tissue which appears paler (c) compared to normal leaves grown in lower light (a). (b), (d) Scale bar = 30 μm. (From Oikawa K *et al.* (2003). *Chloroplast unusual positioning 1* is essential for proper chloroplast positioning. *Plant Cell* 15, 2805–2815. © The American Society of Plant Biologists and reproduced with permission.)

avoidance response. These movements are most clearly seen in the palisade mesophyll cells of dicotyledonous leaves (Fig. 8.7). Movement of chloroplasts in this way can actually be seen with the naked eye on leaves, when different areas of the leaf see a large difference in light intensity.

In *Arabidopsis* leaves, this can be shown very effectively, when a stencil is laid on the leaf and bright light shone on it. This causes a pattern to become visible on the leaf as a result of chloroplasts relocating to the side walls of the mesophyll cells (Fig. 8.7). Such techniques have formed the basis for mutant screens in *Arabidopsis* to identify the genes involved in these two photo-relocation responses by chloroplasts.

The most effective wavelengths of light for inducing chloroplast movement is the blue part of the spectrum and, by characterising *Arabidopsis* mutants unable to show these chloroplast movement responses, a family of proteins that perceive blue light in these responses has been identified. These blue-light photoreceptors are called phototropins, and the genome of *Arabidopsis* contains two phototropin genes, *PHOT1* and *PHOT2*, which function in chloroplast movement responses, but also function in blue-light-induced opening of stomata and phototropic growth responses. Phototropin proteins contain a kinase domain towards the C-terminus and two LOV domains towards the N-terminus. LOV (*l*ight, *o*xygen or *v*oltage sensing) domains are crucial to the blue-light sensing and each domain binds a flavin mononucleotide, which stimulates the kinase activity of the phototropin protein. The two PHOT proteins appear to have overlapping roles in facilitating the two types of chloroplast movement. At low levels of blue light, PHOT1 activates the chloroplast accumulation response; under medium levels of blue light, both PHOT1 and PHOT2 facilitate the chloroplast accumulation movement but in high levels of blue light, PHOT2 is the main receptor that drives the chloroplast avoidance response. Although it has been shown that these phototropin proteins are associated with the plasma membrane in the cell, the way in which the signal transduction pathway functions from blue-light absorbance to induce chloroplast movement is not yet known. However, one component of this mechanism is the JAC1 protein, revealed by the *jac1* mutant of *Arabidopsis*, which fails to accumulate plastids under weak blue light. The JAC1 protein contains a J-domain, resembling the clathrin uncoating factor auxilin at its C-terminus.

However, the basic mechanisms by which chloroplasts can move in cells and the tracks along which they travel are better understood. Chloroplasts move by interacting with the actin microfilament cytoskeleton and move along these actin filaments by using myosin motor proteins. In mesophyll cells of *Arabidopsis* leaves, immunochemical staining of the actin microfilaments clearly shows how they are closely associated with the chloroplasts (Fig. 8.8). The way in which chloroplasts bind to the actin microfilaments to hitch a ride was revealed by identifying mutants in

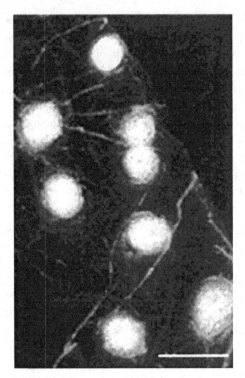

Fig. 8.8. In leaf mesophyll cells in which actin microfilaments are labelled with green fluorescent protein (GFP), the microfilaments can clearly be seen to associate with the envelope membranes of chloroplasts and provide the mechanism by which they can be moved and positioned within cells. Scale bar = 10 μm. (From *Cell Motility and the Cytoskeleton* 44(2) (1999), 110–118. © Wiley-Liss Inc 1996. Reprinted with permission from Wiley-Liss Inc., a subsidiary of John Wiley & Sons Inc.)

which chloroplast relocation, according to the photoenvironment, did not occur. The *Ch*loroplast *U*nusual *P*ositioning 1 (*chup1*) mutant of *Arabidopsis* revealed the CHUP1 protein, which has a distinct structure, suggesting it links the plastid to the microfilaments. The CHUP1 protein contains a hydrophobic region, which locates the protein in the plastid envelope membrane and also has an actin-binding domain which enables binding to actin proteins. It also contains a PRM domain which may well function in polymerising actin monomer subunits into microfilaments, possibly on the chloroplast envelope surface. Thus the CHUP1 protein is probably central to enabling association of chloroplasts with the micro-filament network. Thus three proteins which function in the sensing of light and the subsequent movement of chloroplasts have been identified from *Arabidopsis* mutants, but exactly how they fit together into a signal transduction pathway is yet to be discovered.

It is clear that the capability of chloroplasts to move within the cell's cytoplasm enhances the potential for maximising photosynthetic capacity by optimising light absorbance and to minimise photodamage in high light. Indeed growing plants of the *chup1* mutant in high light causes severe photodamage and bleaching after only 24 hours, simply because the chloroplasts are unable to relocate by using the avoidance mechanism in this mutant. However, when one observes chloroplast distribution in leaf cells grown under moderate light levels, an additional feature is observed, namely that chloroplasts tend to locate on cell surfaces that are exposed to airspaces within the leaf's internal architecture and tend not to reside on cell surfaces that are joined to neighbouring cells. This may well be a feature to maximise the uptake of carbon dioxide from the intercellular airspaces. Such movement behaviour by chloroplasts also explains why chloroplasts undergo such extensive division in leaf mesophyll cells to give rise to large populations of small, individual chloroplasts rather than sufficing with a few giant chloroplasts. Indeed, in mutants such as *arc6*, where there are only a few giant chloroplasts per cell, the plants grow quite normally in moderate light conditions, but the enlarged size of these plastids makes relocation in the cell due to the changing photoenvironment difficult. Thus plastid division in such mesophyll cells has probably evolved to enable large populations of small chloroplasts to develop, each of which has the ability to optimise its cellular position in relation to its photoenvironment.

Plastid differentiation and interconversions

Plastids are unique in plants in that they are the only organelle in plant cells that can take up different forms, the structures of which was considered in Chapter 2. This process is called differentiation and, in most cases in plants growing in normal conditions, is the end of the road for the plastid and it functions according to its differentiated state until it senesces. However, most if not all types of plastid do have the ability to redifferentiate into another type and many of these redifferentiation events occur during different developmental processes in plants. In fact, it is best to regard plastids as a highly dynamic system, with the ability to take on characteristics of a different plastid types either transiently during plant development or to fully redifferentiate into a new type.

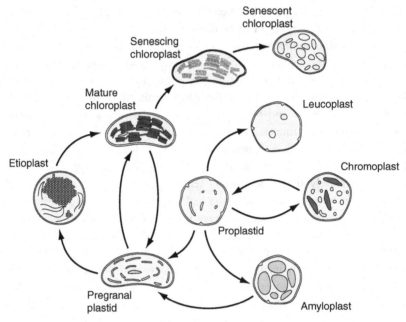

Fig. 8.9. A general scheme which shows how the different types of plastids can interconvert among the different types that occur in different types of tissues in plants. Although specific pathways of redifferentiation are arrowed, it is likely that, within the wide range of cell types found in tissues of higher plants and the vast array of developmental processes that occur, most plastids if not all can differentiate into any other type.

Redifferentiation occurs either as part of a predicted developmental process, such as fruit ripening or as a result of a change in an environmental signal or the presence of a hormone. A scheme of plastid interconversions is shown in Fig. 8.9. All plastid types appear able to form from the progenitor plastid, the proplastid. Many pathways of redifferentiation between plastid types have been studied and are arrowed in Fig. 8.9. However, considering the complexity of higher plants and the relatively few that have been studied in detail, it seems likely that all possible interconversions between different plastid types probably occur in nature to some extent. It is normally considered that the differentiation of chloroplasts from proplastids during the formation of green tissues in plants is the default pathway for plastid differentiation and that the primary differentiation of other plastid types from proplastids or the subsequent redifferentiation of chloroplasts into a different plastid type are processes that have evolved during the evolution of plants.

There are several classic, well-studied redifferentiation pathways where one plastid type changes into another. The ripening of fleshy fruits is a good example, where chloroplasts redifferentiate into chromoplasts as part of the ripening process. This occurs in tomatoes, peppers, citrus fruits and some Cucurbits including pumpkins. Tissues of these structures are green when immature but are pigmented orange, yellow or red when mature. There are subtle differences between the ripening of tomatoes and peppers and the ripening of citrus fruits and pumpkins, since in the former the redifferentation of chloroplasts into chromoplasts is terminal, whereas in citrus and pumpkins the pigmented chromoplasts can redifferentiate back into green chloroplasts, a process that appears to be controlled by the plant hormone, gibberellin. Another well-documented trait is the greening of non-green plant tissues when exposed to light. The classic example here is the greening of potato tubers when exposed to light, which triggers the redifferentiation of amyloplasts in the tuber cells into green chloroplasts. This trait has serious commercial implications in the storage of potato tubers, since green potatoes are less desirable and the green tissues accumulate a toxic alkaloid called solanin. Plastid redifferentiation also occurs during development of petals in flowers of some plant species. Many petals contain green chloroplasts when they are small and within the unexpanded flower bud, but during the latter stages of petal development, the chloroplasts can redifferentiate into chromoplasts, yielding colour to the petal or can redifferentiate into colourless leucoplasts, which lose their chlorophyll and make the petal white. In petals that have pigmented areas on a white background, these two contrasting redifferentiation pathways could occur in neighbouring cells. Another well-studied plastid differentiation pathway, mentioned previously, is the light-driven conversion of etioplasts into chloroplasts as etiolated tissues green up on exposure to light. In a more artificial system, leaves that are senescent and contain gerontoplasts can be induced to regreen and the gerontoplasts redifferentiate back into chloroplasts, by the application of hormones, namely cytokinin.

Although these interconversions between different plastid types have been well described, the understanding of the molecular processes that control and underlie them is poor. It is clear that a subset of nuclear gene products represent each type of plastid, and an analysis of the proteome of different plastid types clearly shows that they differ significantly in their spectrum of proteins. However, it is less clear what are the major control factors which determine which sort of plastid is made in which cell type. However, most plastid types contain some chloroplast features, namely residual thylakoid membrane and low levels of expression of genes encoding chloroplast proteins, reflecting the fact that the chloroplast is normally considered the default pathway for plastid development.

Whether there are critical nuclear genes that are required for the default pathway of chloroplast differentiation is unclear, although efforts have been made to identify mutants that reveal master controllers of chloroplast biogenesis. Unfortunately, mutating any of a wide variety of plastid related genes results in perturbed chloroplast development. Identifying core genes that facilitate chloroplast differentiation is an ongoing effort that should reveal the key components that enable chloroplasts to develop in leaf cells. Amongst the best candidates currently are *GLK* genes, which are conserved in all land plants and encode nuclear transcription factors, which when mutated cause serious perturbations in the development of the chloroplast and the thylakoid membrane.

A subtle and important differentiation pathway that occurs in plants carrying out a C_4 type of photosynthesis is the different pathway of chloroplast differentiation in neighbouring cells in leaves. In C_4 plants, typified by species such as maize (*Zea mays*) and sugar cane (*Saccharum* sp.), two distinct types of photosynthetic cells develop in the leaf: those cells surrounding the vascular bundles of the leaf are termed bundle sheath cells and those photosynthetic cells resident throughout the body of the leaf are mesophyll cells (Fig. 8.10). This type of leaf anatomy is generally referred to as Kranz anatomy. Chloroplasts develop differently between these two cell types, since a biochemical mechanism exists, which fixes carbon dioxide using the enzyme phosphoenol pyruvate carboxylase (PEP carboxylase) in the cytoplasm of the mesophyll cells. This facilitates the re-release of carbon dioxide in the bundle sheath cells, where it is refixed by the enzyme RUBISCO in the bundle sheath cell chloroplasts. The resulting elevated concentration of carbon dioxide in the bundle sheath cell chloroplasts overcomes problems of photorespiration caused by the oxygenase activity of RUBISCO, especially at higher temperatures, in which C_4 plants grow preferentially. There are clear differences between chloroplasts in the mesophyll cells compared with the bundle sheath cells, the most obvious being that thylakoid membranes in bundle sheath cell chloroplasts are largely agranal in that they lack thylakoid stacks. Thylakoid membranes from bundle sheath chloroplasts are devoid or depleted in activity of photosystem II, even though most of the PSII proteins are present but cannot evolve oxygen via the oxygen-evolving complex as normally occurs in C_3-type chloroplasts (see Chapter 4). There is clear repression of gene expression for RUBISCO genes in the mesophyll cells, resulting in a total lack of RUBISCO enzyme in these cells.

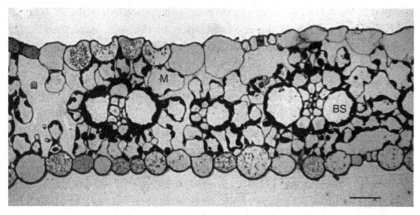

Fig. 8.10. Two distinct types of chloroplast-containing cells are present in leaves carrying out C_4 photosynthesis. This section through a maize leaf shows a distinct ring of cells around the vascular bundles, termed bundle sheath cells (BS), in which the darkly stained chloroplasts take up a peripheral position in the cells, farthest from the vascular tissue. The other chloroplast-containing cells are the mesophyll cells (M), which fill up the bulk of the leaf's internal architecture. Scale bar $= 25$ μm. (Image courtesy of Jane Langdale, University of Oxford, UK.)

A proteomic analysis of the two types of chloroplasts show clear differences in other aspects of their metabolism, with genes involved in lipid synthesis, nitrogen import and tetrapyrrole and isoprenoid biosynthesis preferentially expressed in mesophyll cells whereas in bundle sheath cells there is preferential expression of genes involved in starch synthesis and sulphur metabolism associated with the chloroplast.

Chloroplasts in these two cell types also position themselves differently within the cell. In the mesophyll cells, the plastids tend to be generally distributed but in the bundle sheath cells, the chloroplasts are normally located at the periphery of the cells in a centrifugal manner, farthest from the vascular cells and nearest to the mesophyll cells (Fig. 8.10), a positioning that facilitates the transfer of molecules between the two cell types and their chloroplasts. Thus chloroplasts clearly differentiate into two distinct types in these two cell types, which are mostly adjacent to each other. At least two factors appear to control this cell-specific chloroplast differentiation. Firstly, light is a crucial factor in causing differential expression of genes encoding the various C_3 and C_4 enzymes between the bundle sheath and mesophyll cells, since in etiolated leaves grown in the dark, RUBISCO enzyme is synthesised in all cells but in light-grown leaves is only found in bundle sheath cells. Thus, light causes the repression of RUBISCO gene expression in mesophyll cells. A second factor is the position of the cell in relation to the vein, since those cells close to the veins show repression of RUBISCO synthesis but, in instances where veins are widely spaced and there may be up to 20 or more mesophyll cells between veins, those particular mesophyll cells most distant from a vein take on C_3 characteristics and synthesise RUBISCO. So, somehow, light and cell position relative to a vein interact to induce this dimorphic differentiation of chloroplasts. Furthermore, there is a difference in the expression of two different *GLK* genes between mesophyll and bundle sheath cells in maize. *ZmGlk1* is expressed primarily in the mesophyll cells and *G2* is expressed primarily in the bundle sheath cells, suggesting that this difference in *GLK* expression may be an underlying control for chloroplast differentiation in leaves of C_4 plants.

Until recently it has always been thought that a Kranz type of anatomy was crucial for enabling differentiation of dimorphic types of chloroplasts in different cell types, enabling C_4 photosynthesis to occur. However, examples of C_4 photosynthesis being carried out in single leaf cells has been found in leaves of distinct species of the family *Chenopodiaceae*. In these species, ultrastructurally distinct types of chloroplasts are partitioned within a single cell into two biochemically defined compartments, thereby mimicking the spatial separation of mesophyll and bundle sheath cells within one cell. In one species, the two compartments are maintained as the peripheral cytoplasm and a central cytoplasmic space joined by cytoplasmic strands, each containing a distinct type of chloroplast, whereas in a different species, the separation is essentially maintained by the cytoplasm at the top and bottom of long elongate cells. Consequently, the cell contains two distinct types of chloroplasts, but in a different position within it. These two distinct chloroplast compartments are maintained in their defined positions by close interaction with the microfilaments and microtubules of the cytoskeleton which ensures that, along with other cellular organelles, the spatial positioning is maintained.

Plastids in root cells

Leucoplasts are essentially unpigmented plastids, which are found in many different types of cells within plant tissues and are essentially the plastid type referred to as non-green plastids. A major class of leucoplasts are those found in the cells of roots and, in terms of total plastid numbers within the body of a plant, probably comprise a significant component of

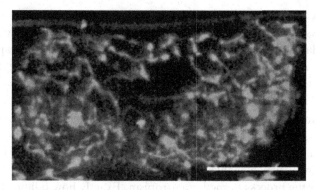

Fig. 8.11. Targeting GFP to plastids in tobacco roots, in which mycorrhizal arbuscules formed in the root cortex cells reveal an extensive network of plastid bodies and stromules covering the fungal arbscules. Bar = 25 μm. (From Fester T, Strack D, Hause B (2001). *Planta* 213 864 with kind permission from Springer Science + Business Media.)

the total. Root plastids play a fundamental role in root cell biochemistry and metabolism and carry out much of the basic metabolism as described in Chapter 7. They are normally unpigmented (see Fig. 2.10), but can be visualised easily when green fluorescent protein is transgenically targeted to the plastid (see Fig. 8.6, Fig. 2.13). They are highly variable in morphology and show extensive stromules, which could well be considered as the most extreme part of a highly heterogeneous morphology.

Although the main role of root plastids is in their major contribution to root cell metabolism, they have been shown to have a distinct role in enabling root cells to interact with symbiotic fungi and bacteria. In cells of the root cortex, which are invaded by fungi and are transformed into arbuscules, the plastids in these cells develop large stromule networks, which interact with the extensive surface of the fungal hyphae (Fig. 8.11). In addition to this stromule network, metabolic and transcript profiling shows that the metabolic activity of the cell's plastids is up-regulated dramatically, which provides an array of metabolites that are vital to setting up a symbiotic interaction with the invading symbiont. Two plastid envelope membrane proteins, CASTOR and POLLUX, control the first entrance of microbes into the cell at the earliest stage of establishing a symbiotic relationship, such as that which occurs during the establishment of nitrogen fixing root nodules. It is interesting that a long-term endosymbiont of the plant cell, the plastid, now plays a role in admitting more short-term endosymbionts to the plant cell in these special situations.

An excellent example of cell-specific plastid differentiation is that of amyloplast differentiation in the columella cells in the tips of roots. Columella cells form ahead of the meristem, behind the root cap at the root tip and, within them specifically, amyloplasts differentiate and can easily be visualised when stained with iodine (see Fig. 2.10). These specialised amyloplasts are termed statoliths and are involved in the root's perception of gravity. Statoliths fall to the bottom of the cell under the influence of gravity, since they contain dense starch and are less buoyant within the cytoplasm than other organelles. This process initiates a signal transduction pathway, which enables root growth downwards towards the gravity vector, in the process termed gravitropism. Exactly how falling statoliths initiate a signal that causes differential cell expansion of the root, thereby altering its direction of elongation is unclear, although the fallen statoliths probably interact with the endoplasmic reticulum in some way and the resulting signal transduction pathway involves a redistribution of the hormone auxin between the upper and lower sides of the root. Little is known also about what initiates amyloplast differentiation from root cell leucoplasts in the collumella cells, but not in the surrounding cells. Interestingly, changing patterns of auxin distribution around the root tip can initiate statoliths production in other cell types and, in mutants of *Arabidopsis* that affect auxin transport, the distribution of statoliths in cells is altered. Thus auxin may be a factor involved in enabling cell-specific amyloplast differentiation in this way. In addition, this cell-specific differentiation pathway is only transient since subsequently, columella cells are pushed farther towards the root tip and become root cap cells, in which amyloplasts revert to leucoplasts, and are eventually sloughed off the root into the soil.

Plastids as sources of developmental signals

In addition to the complexity of the plastid's interaction with the cell and its nucleus, as has been outlined in the preceding chapters, there is a body of evidence that suggests that the plastid can direct the differentiation pathway of the cell itself. This has been shown primarily in the differentiation of elongated palisade mesophyll cells in the leaf, which differentiate as one or two layers of cells beneath the upper leaf epidermis. Thus, in dicotyledonous leaves there are two clearly defined types of mesophyll cell: palisade mesophyll cells towards the top and spongy mesophyll cells towards the underside of the leaf, internal to the lower epidermis.

Evidence that functional chloroplasts might direct the differentiation of palisade mesophyll cells comes largely from examining a variety of mutants in which chloroplast function is perturbed, for a variety of different reasons, and the resulting internal leaf architecture is also altered. In all of these cases, the loss of functional chloroplasts during leaf development causes a failure of the palisade mesophyll cell layers to differentiate correctly, yet spongy mesophyll cells develop as normal. A clear demonstration of this phenomenon occurs in the variegated mutant of *Arabidopsis* called *immutans*. This mutant is impaired in carotenoid biosynthesis and gives rise to a variegated leaf phenotype, which contains sectors that are normal green and those that are pale. In the pale sectors, in which plastid development is impaired, the development of the palisade mesophyll cells is severely perturbed and this is clearly shown at the boundaries between sectors. Thus, it is hypothesised that some aspect of plastid function or a plastid-derived signal induces palisade mesophyll differentiation in dicotyledonous leaves.

9

Plastid transformation and biotechnology

The revolution in genetics and molecular biology that occurred towards the end of the twentieth century inspired enormous progress in understanding many aspects of the molecular control of development of different organisms. It has also aided our understanding of how information contained in their respective genomes gives rise to large populations of different proteins, termed the proteome. In turn, these proteins interact with metabolites enabling the organism to develop its specific phenotype. Moreover, the use of molecular genetic techniques enabled genetic systems to be altered artificially in order to exploit aspects of molecular synthesis or developmental biology for improvement of agricultural species of plants and animals.

Early on in this revolution, plastids were seen as an attractive system in which plant biotechnology could be performed, since they have several properties that are advantageous in this area. Also, as we have seen throughout this book, plastids carry out many vital processes in plant cells, which have the potential to be manipulated for improvement or increased efficiency, leading to crop plants with more optimised phenotypes for specific environments. Because plastids are organelles, which define a distinct compartment within the plant cell, sequestering novel molecules into the plastid compartment presents a major advantage compared with allowing accumulation of novel molecules in the cytosol where they could be toxic. Since the relative proportion of the plastid compartment size compared to the cytoplasmic·volume is fairly high, the plastid represents a significant compartment in which novel molecules can be accumulated.

Introducing novel proteins into a plant requires the establishment of a transgenic plant expressing the gene for the novel protein. There are different methods by which novel molecules can be accumulated in the plastid compartment in a transgenic plant. Firstly, a transgene can be introduced into the nuclear genome, which expresses a novel protein possessing a plastid transit sequence, which enables the novel protein to be imported into the plastid stroma. Surprisingly, the Tic Toc import systems of the plastid envelope appear to be easily tricked into importing novel proteins into the plastid, and a wide variety of proteins have been targeted to plastids in different transgenic plants. One already described in this book is green fluorescent protein (GFP), a protein from the jellyfish *Aequorea victoria*, which is easily targeted to plastids enabling stromules to be observed (see Chapter 8), and has been widely used, along with other plastid-targeted fluorescent proteins, to study a variety of aspects of plastid cell biology.

Plastid transformation technology

A key technology that has been used to understand how organisms function at the molecular level has been the insertion of foreign DNA sequences into nuclear genomes in order to modify how host genes are expressed, or to express novel genes in molecular environments where they would not previously have been found. Such transgenic technologies have been exploited far and wide in all organisms, including plants. Since the plastid contains its own genome, as was considered in detail in Chapter 3, the possibility of inserting foreign DNA directly into the plastid genome was a distinct goal from the earliest days of plant genetic manipulation research, but turned out to be a complex and difficult goal to achieve for several years. Traditionally, the most commonly used way of introducing DNA into the nuclear genome of plants exploits the soil bacterium, *Agrobacterium tumifaciens*, which has a natural system as part of its plant infection process, in which a piece of DNA termed T-DNA in part of its Ti-plasmid is inserted into the chromosomal DNA in the plant's nuclear genome. The development of this technology has enabled stable nuclear transformations in plants to be made using *Agrobacterium* transformation in a routine manner. Another method for plant nuclear transformation has used a more physical approach; that of coating small tungsten pellets with plasmids of vector DNA and shooting them through plant tissues. The DNA, which sloughs off inside cells, enters the nucleus and becomes incorporated into the nuclear chromosomes. This approach is termed biolistics and has been used successfully in transforming the nucleus of a wide variety of plants. Thus, *Agrobacterium* transformation

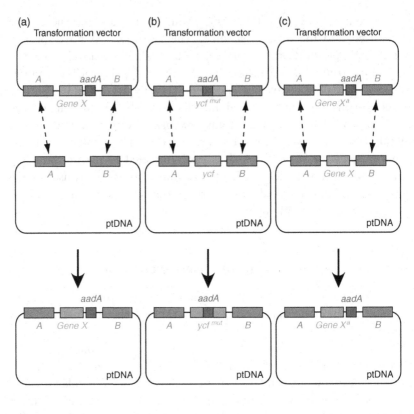

Biotechnology
Integration of transgenes
(*X* = gene of interest, GOI)

Reverse genetics
Insertional mutagenesis
Deletional mutagenesis

Gene replacement
(*X* = endogenous gene;
X^a = allele of gene *X*)

Test of expression elements
Promoters, UTRs, fusion tags
(*X* = reporter gene or GOI)

Reverse genetics
Introduction of point mutations

Fig. 9.1. Plastid DNA (ptDNA) in the plastid genome can be transformed by vector DNA introduced into the plastid by biolistics of a transformation vector. The design of the vector DNA will result in different types of transformation. (a) For the introduction of a novel gene of interest (gene *X*). The gene is placed alongside a selectable marker gene (*aadA*), conferring resistance to spectinomycin, and flanked by two sequences (A, B), which have homologous sequences in the plastid DNA. The A and B sequences recombine inserting gene *X* and *aadA* into the plastid DNA. Sequences A and B are likely to be sequences in an intergenic region of plastid DNA to minimise changes to plastid DNA gene function. (b) In this approach, the selectable marker gene (*aadA*) is inserted in the middle of a plastid gene (*ycf^{mut}*) and the recombination sequences are present either side of the *ycf* gene in the plastid DNA. As a result of recombination,

and biolistic transformation were the two strategies that were used to try and transform plastids at the outset of plastid biotechnology.

Despite some initial suggestions that *Agrobacterium* transformation of plastids might be feasible, the entire transformation technologies that have been developed to transform plastids have used biolistics. This technique uses pieces of plant tissue or cell suspensions, which are shot with a biolistics gun, and vector DNA on the pellets is incorporated into the plastid genome of a few plastids in the cells through which the tungsten particles pass or become lodged. Design of the vector DNA sequence is critical to successful plastid transformation and the DNA sequence to be introduced into the plastid DNA requires several crucial parts in order to optimise plastid transformation efficiency and the selection of transformants. The gene sequence to be introduced into the plastid genome is flanked by two DNA sequences, which are homologous to neighbouring DNA sequences on the host plastid genome (Fig. 9.1). The actual transformation process into the plastid DNA exploits the process of homologous recombination, which is carried out in plastids by a recombinase enzyme recA. This enzyme is nuclear-encoded and similar in structure to the recombinase enzyme found in bacteria. Normally, in the plastid it functions to recombine different parts of the plastid genome, which are homologous, most notably the inverted repeat sequence, which can generate two different circular forms of the plastid DNA molecule and confers stability on the genome (see Chapter 3). However, in the artificial situation of plastid transformation, the recA enzyme recognises vector sequences homologous to those present in the host plastid DNA. Thus, by homologous recombination between the sequence on the plastid genome and that on the vector DNA sequences, the transgene is introduced into the plastid genome (Fig. 9.1). Consequently,

Caption for Fig. 9.1. (cont.)
the plastid *ycf* gene is replaced by the version with *aadA* in the middle, resulting in an insertion mutation of *ycf*. Such approaches can be used as a reverse genetics approach to investigate plastid gene function by insertional or deletion mutagenesis. (c) In this approach, a gene X^a, carrying a point mutation in the *X* gene sequence is introduced alongside the *aadA* selectable marker and replaces completely the normal plastid encoded gene *X*. This approach can be used for gene replacement and the introduction of alterations in gene sequence to examine gene function within the plastid genome. (Redrawn from Bock R, Plastid biotechnology: prospects for herbicide and insect resistance, metabolic engineering and molecular farming. *Current Opinions in Biotechnology* 18, 100–106. © Elsevier (2007).)

plastid transformation vectors can be designed to target transgenes to specific sites in the plastid genome, chosen according to the nature of the experiment and the outcome desired. Thus, the only sequences introduced into the plastid genome flanked by the recombination sites are the gene of interest and a gene expressing a selectable marker. If the experiment is designed simply to introduce a transgene expressing a novel protein into the plastid, then homologous integration sites in the plastid genome would be chosen to be in intergenic regions between functional genes to minimise disruption to plastid function (Fig. 9.1).

Using a selectable marker gene is crucial in selection for those plastid DNA molecules that have been transformed, such that they replicate preferentially over those that are not transformed when the selective agent is present in the plastid. The most commonly used selectable marker for transgenic plastids is the *aadA* gene, which confers resistance to the antibiotic spectinomycin. The *aadA* gene encodes the enzyme aminoglycoside 3''-adenylyltransferase (AAD), which inactivates spectinomycin. Following biolistics, some plastids take up the transformation vector, which in turn confers resistance to the antibiotic spectinomycin. Cells that contain such transformed plastids will grow preferentially under selection for spectinomycin resistance. There is a complexity here in that tissues contain large numbers of plastid genomes in their population of plastids and several rounds of selection and plant regeneration are required in order to achieve homoplastomic plants, the state in which all of the genomes in all plastids within a plant contain the transgene. Thus, replication of transgenic plastid DNA molecules, the division of plastids containing transformed plastid DNA and the division of cells containing them all contribute towards achieving a state of transgenic homoplasmy in a tissue. Experimentally, this is achieved by regenerating whole plants from biolistically shot leaf tissue through the process of tissue culture. Successful plastid transformation in this way was achieved first using the model system tobacco (*Nicotianum tabacum*) and has become the species in which the vast majority of plastid transformation experimentation has been carried out. Generation of new shoots from biolistically shot leaves in tobacco tissue culture involves dedifferentiation of leaf chloroplasts into proplastids and the generation of new meristems, followed by the induction of shoots and roots in tissue culture, and finally the generation of a new tobacco plant. Figure 9.2 shows how tungsten pellets containing vector DNA can interact with plastid DNA in nucleoids and how transgenic plastids can be sorted out by selection and plastid DNA replication and plastid division leading to the homo-plastomic state in all cells in the regenerated tobacco plant.

There remains a problem with the implementation of this plastid transformation technology in a wide range of species, including important crop species. To date, plastid transformation is only a routine procedure in tobacco. More recently, plastid transformation has been achieved in several other crop species including tomato, potato, soybean, cotton and poplar. It would appear that, with future developments, plastid transformation will potentially become routine in most crop plants, although plastid transformation of cereals, including important crop plants such as wheat and maize, has yet to be achieved. Although a leaf biolistics approach to transformation works well in tobacco, this may not be the best approach in all species. An alternative approach of using cell cultures for biolistic transformation and subsequent production of somatic embryos has been developed, and has been shown to work in cultures of carrot, cotton and soybean. In this approach, cell suspension cultures are established and either these or lumps of undifferentiated callus tissues are bombarded with tungsten pellets, using the same equipment as in leaf biolistics. A thorough understanding of the tissue culture behaviour of the species is then required in order to initiate cells dividing and differentiating, giving rise to somatic embryos, each of which can then generate a whole plant.

In terms of general plant biotechnology, the transformation of plastids and sequestering of a novel gene product within the plastid compartment has a range of advantages over that of transforming the nuclear genome and having the gene product accumulate in the cytoplasm. Firstly, the high copy number of the genome in plastids means that the transgene will be present in the cell in significantly higher copy number than if the nucleus alone were transformed. In a stable nuclear transformation, there will normally be two copies of the transgene per cell, one allele on each of the pair of homologous nuclear chromosomes. This compares with up to 10 000 copies of the plastid genome per cell in tobacco and thus, in a homoplastomic plant, there would be up to 10 000 copies of the transgene per cell. Thus the potential for producing large amounts of any novel transgene product is greatly enhanced by using plastid transformation compared with nuclear transformation. Secondly, the ability to target insertions into the plastid genome precisely via directed homologous recombination means that the position of the transgene is known exactly, which contrasts dramatically with nuclear transformation using *Agrobacterium*, in which DNA inserts are essentially random throughout the

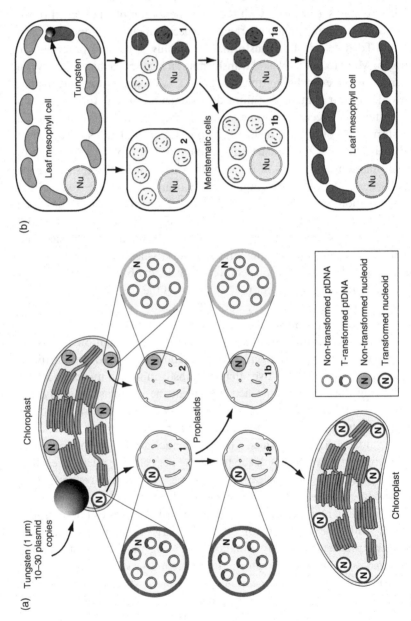

Fig. 9.2.

genome and often complex and rearranged. Thirdly, expression of a transgene on the plastid genome does not appear to be compromised by epigenetic effects or by mechanisms of gene silencing, which cause significant problems in nuclear transformation causing novel transgenes to be silenced in a variety of different ways. Fourthly, since the plastid has essentially a prokaryotic–type genome, it has the ability to receive more than one transgene at a time in terms of whole prokaryotic-type operon units. Thus in theory, an operon from a bacterial genome containing several genes could be inserted in its entirety into the plastid genome and be successfully transcribed and translated into its constituent gene products. All of these advantages make the use of plastid transformation in plant biotechnology an attractive proposition. There are possible disadvantages to using transplastomic technology to produce novel proteins, one being the fact that the plastids appear to have few systems to modify proteins post-translationally, a role that is normally carried out in the cell by the endomembrane system of endoplasmic reticulum and Golgi bodies. In many cases this may not be essential, but in cases where a protein needs modifying by, for example, glycosylation and the addition of sugar moieties to make it active, transplastomic approaches may not be the best.

Another area in which plastid transformation has a unique advantage is in the environmental containment of transgenes. In the majority of Angiosperms, during the development of pollen, proplastids are lost even though they are present at the earliest stages, and their DNA is degraded (see Chapter 2). This means that, in mature pollen, there are no plastids or

Caption for Fig. 9.2.
The process of plastid transformation by biolistics requires two types of sorting out of transformed ptDNA to yield homoplastomic cells. (a) Vector DNA from the tungsten pellet lodged in the plastid transforms some ptDNA molecules in a nucleoid, which then replicates under selection conditions (1a) during the process of plastid dedifferentiation to proplastids during tissue culture. Replication of nucleoids preferentially replicates transformed plastid DNA molecules leading to homoplastomic nucleoids in proplastids. Nucleoids containing no transformed plastid DNA (1b) are out-replicated under selection conditions. The differentiation of chloroplasts during shoot induction in tissue culture leads to chloroplasts which are homoplastomic for the transformed plastid DNA in all nucleoids. (b) A cell containing a homoplastomic plastid will preferentially replicate it under selection conditions such that the result of plastid division and cell division (1, 1a, 1b, 2) will give rise to cells which are homoplastomic in all chloroplasts. (Redrawn from Maliga P (2004). *Annual Review of Plant Biology* 55, 289–313. Reprinted with permission from the *Annual Review of Plant Biology*, 55 © 2004 by Annual Reviews www.annualreviews.org.)

plastid DNA. Consequently, a plastid transgene in a transplastomic plant will not be dispersed into the environment via pollen dispersal. Since problems with widespread transgene dispersal was considered a major problem in the use of transgenic plans in agriculture, the insertion of transgenes into the plastid represents a neat way of transgene containment. Even so, a general problem with growing transgenic plants is the release of antibiotic resistance genes into the environment from any transgenic tissues, since it is considered that these antibiotic resistance genes could pass to bacteria in the environment and especially those in the soil, and generate widespread antibiotic resistance. A way around this problem is to excise the antibiotic resistance marker from the transgene, once the transplastomic plant is homoplastomic and stable, such that the final transgenic plant lacks any antibiotic resistance. This can be done using a nuclear-encoded enzyme that recognises specific excision sequences within the integrated piece of vector DNA placed either side of the selectable marker sequence and recombines the two sites, excising the selectable marker sequence in the process. These are different recombination sites to those used by recA for the initial integration event. A commonly used system which allows selectable marker excision in transplastomic plants is the Cre-Lox system, which uses a recombinase from a bacteriophage inserted transgenically into the plant nucleus. Cre-Lox recognises specific recombination sequences, which are built either side of the selectable marker gene sequence in the vector designed for plastid transformation. Once a homoplastomic plant has been established and contains a selectable marker gene along with the gene of interest, a nuclear-encoded Cre-Lox together with a plastid targeting sequence to enable it to be imported into the plastid, can be introduced by genetically crossing the transplastomic line with a nuclear transgenic containing Cre-Lox. In the resultant plants, the nuclear-encoded Cre-Lox will be expressed, imported into the plastids and will recombine its two target sequences in the vector sequence in the plastid DNA, thereby excising the selectable marker gene in all plastids in the plant. This will result in a transplastomic plant with no antibiotic resistance marker remaining in any of its cells.

A further strategy which may become fruitful in the future is the design and introduction of circular DNA molecules into plastids, which are self-replicating because they contain an origin of replication sequence that enables them to proliferate independently in the plastid. These vectors are termed shuttle vectors and have been successfully incorporated into plastids, although their long-term viability appears limited without continuous selection.

Plastid transformation and the functional analysis of plastid genes

Using homologous flanking sequences, which reside within a single gene, it is possible to insert the selectable marker gene or any other sequence of DNA into a given plastid gene and thereby make an insertional or a deletion mutation in that gene (Fig. 9.1(b)). Such strategies have been used widely to try and identify the function of specific plastid-encoded genes by knocking them out in this manner and looking for a resultant phenotype. Also, it is possible to replace specific plastid genes with allelic versions containing point mutations of genetic rearrangements (Fig. 9.1(c)), and also to try and reveal how specific genes or parts thereof function. Such strategies have been used widely and have generated much of the knowledge about plastid gene function and expression discussed earlier in this book.

Plastid transformation to improve crop plant performance

The ability to transform plastids allows different molecular genetic strategies to be developed for addressing various questions about plastid function or plastid biotechnology. The simplest strategy is to introduce a novel gene into the plastid genome and to observe how it is expressed and how its protein product accumulates (Fig. 9.1a). This approach has been used in a variety of different experiments, with the aim of either conferring a novel trait on a plant that is agriculturally important or in accumulating a biologically useful molecule in the plastid, an area often referred to as molecular farming.

In the first of these approaches, conferring resistance to herbicides or insect pests has been addressed using transplastomic technology. The commonly used herbicide, glyphosate, has its mode of action in preventing amino acid biosynthesis in the plastid by inhibiting the key enzyme 5-enol-pyruvyl shikimate-3-phosphate synthase (EPSPS) (see Chapter 7). EPSPS is a nuclear-encoded plastid-targeted protein, which is also found in bacteria and for which many different glyphosate insensitive alleles exist, either naturally occurring or by the introduction of point mutations. The introduction of novel EPSPS sequences into the plastid genome has the potential for increasing the amount of glyphosate-tolerant EPSPS molecules within the plastid and also testing how different alleles and different expression signals affect the levels of the protein. Plants transplastomic for EPSPS can be made, which are highly resistant to glyphosate and accumulate EPSPS to levels around 10% of plant total soluble protein. Interestingly, the gene used in these experiments is the full-length gene, which is normally nuclear-encoded and contains the plastid targeting sequence, as would be expected in a plastid-targeted protein. When this sequence is expressed from within the plastid, in the ptDNA, the protein is transcribed containing a plastid transit sequence, although it is not functionally required. Even though the protein has not been imported into the plastid, but made inside the plastid, the stromal processing peptidase that normally processes imported proteins also processes the EPSPS protein, removing its transit peptide sequence. In another example, transplastomic plants were made containing the *bar* gene, encoding the enzyme phosphoinothricin acetyl transferase (PAT), which confers resistance to the herbicide phosphinothricin. In this case, the PAT enzyme accumulated to levels up to 7% of total leaf soluble protein and conferred full resistance to the herbicide when applied in field conditions.

In relation to making plants resistant to insect pests, transplastomic technology has been highly successful. The crystal toxin proteins (Bt toxin) produced by the bacteria *Bacillus thuringiensis* are toxic to insects and have been a major target for conferring resistance to insect damage in transgenic crop plants. The *cry* genes encoding Bt toxin are prokaryotic in nature and the nature of their codon usage makes efficient expression from plant nuclear genomes complex. However, when expressed transplastomically in the prokaryotic-type environment of the plastid genome, their expression is fine and there is no requirement for tinkering with sequence or codon usage to maximise expression. Such transplastomic tobacco plants are highly toxic to grazing insect larvae. In spectacular fashion, Bt toxin can accumulate up to 45% of total soluble protein in these plants when the *cry* gene formed part of a three-gene operon, identical to that present in the bacterial genome. In this case, one of the other genes contained within the operon acts as a chaperone to assist folding of the Bt toxin protein in the plastid and leading to protein crystallisation within the stroma. The presence of cuboid crystals of the Bt toxin inside the plastid is a dramatic demonstration of the power of transplastomic technology. Thus, this piece of chloroplast biotechnology works well, although high levels of Bt toxin may come with a penalty of retarded plant growth. Obviously, it is crucial in such strategies to optimise recombinant protein levels in order to confer the desired novel plant phenotype but not to produce it excessively, resulting in aberrant plant performance or undue toxicity from the novel protein in the organelle.

It is also feasible to engineer the plastid to make it more tolerant of environmental stresses, such as drought, salinity and temperature. In these conditions, plants can synthesise osmoprotectant molecules, which enhance water uptake and retention by the cell and stabilise macromolecules in the cell, preventing damage to them by high levels of salts, which can arise in plants exposed to these stresses. One osmoprotectant molecule is the disaccharide, trehalose, which has been implicated in protection against these various environmental stresses and has been shown to accumulate in plant cells when they are exposed to extreme temperature, drought or salinity. Trehalose is synthesised by the nuclear-encoded enzyme trehalose phosphate synthase, encoded by the nuclear gene *TPS1*. If this gene is over-expressed transgenically on the nuclear genome, the plants show abnormal pleiotropic effects in their phenotype, which is undesirable but, when expressed from the plastid genome in transplastomic plants, plant growth is normal and trehalose content is increased by 25-fold compared with the wild type. Droughting such plants artificially shows that these transplastomic plants are significantly more drought tolerant than normal wild-type plants. This is an important demonstration of the advantages of sequestering increased levels of protein within the plastid compartment rather than allowing them to remain in the cytosol, where significant changes in the concentration of a given protein often seem to cause undesirable effects on cell and plant function. However, when accumulated within the plastid compartment, this seems to be less often the case.

Another osmoprotectant, which is synthesised by some plants when faced with drought conditions, especially those brought on by salinity stress, is glycine betaine, which is a very effective osmoprotectant and helps to maintain an osmotic balance within the cell. The fact that several important crops do not exhibit this drought physiology response raises the possibility of enabling synthesis of glycine betaine in crop plants to enhance resistance to drought and salinity stress. Glycine betaine is synthesised by two enzymes, which are normally located in the plastid: choline monooxygenase (CMO) and betaine aldehyde dehydrogenase (BADH). Insertion of the *BADH* gene into the plastid genome in transplastomic carrot (*Daucus carota*) was achieved by biolistically transforming carrot cell suspension cultures, which were then induced to develop somatic embryos, subsequently giving rise to mature homoplastomic carrot plants expressing a *BADH* gene on the carrot plastid genome. Using this approach, transplastomic carrot plants were shown to have greatly enhanced levels of glycine betaine and show tolerance of sodium chloride concentrations up to 400 mM, which represents a serious salinity stress.

A further possibility in improving plant resistance to environmental stress comes in enhancing the plant's fight against pathogens. Although complex molecular systems exist, which enable plants to defend themselves against pathogen attack, diseases cause very major losses in crops worldwide annually. The possibility of synthesising antimicrobial resistance within plastids was achieved by using an analogue of maganin-2, which is a helical antimicrobial protein which gives some protection against bacteria and fungi by binding to their cell membranes and causing cell lysis. The analogue is called MSI-99. When MSI-99 is expressed on the plastid genome in tobacco leaves, application of various pathogens to the leaves does not result in leaf infection and showed good levels of pathogen resistance. Once again, this was a much better result than when the genes were expressed on the nuclear genome. This raises a serious possibility of producing transplastomic crop plants that exhibit pathogen resistance.

Another novel aspect of plastid biology, which can be exploited in transplastomic technology, is that of heavy metal damage and efforts to generate plants which are resistant to heavy metals in the general process of phytoremediation; that is using plants to clean up polluted environments. Mercury is a highly toxic heavy metal, whose main target for damage in the cell is the plastid. Consequently, engineering genes into the plastid genome that can detoxify mercury represents an attractive option for phytoremediation. The bacterial operon containing the genes *merA* and *merB* encode the enzymes mercuric ion reductase and organomercurial lyase, respectively. Between them, these two enzymes convert mercury into a less toxic form and offer a distinct possibility for mercury phytoremediation. Producing homoplastomic tobacco plants containing the *merA/merB* operon on the plastid genome showed this to be the case. When such plants were challenged with the toxic compound phenylmercuric acetate (PMA) in their growth medium, they showed significantly better biomass accumulation than wild-type plants. Thus, this approach to phytoremediation exploiting transplastomic technology appears a valid option.

Plastid transformation and metabolic engineering

A more subtle approach to plastid genetic engineering is to consider the central role that the plastid plays in the biochemical metabolism of the plant cell and to identify aspects of plastid metabolism that can be altered using plastid transformation technology. These can include efforts to change the relative flow through different parts of metabolic pathways, to change the balance of final molecular products or to try and improve the efficiency of a plastid metabolic process. Foremost in this latter case is a consideration of photosynthesis and whether different aspects of it could be improved through plastid engineering. A major focus for genetic manipulation via transplastomic technology has been the large subunit of the carbon fixation enzyme RUBISCO (see Chapter 4), the gene for which is normally encoded on the plastid genome, with the gene for the small subunit nuclear-encoded. The functional holoenzyme contains eight large subunits and eight small subunits. There are two aspects of the RUBISCO enzyme that can potentially be addressed by replacing the large subunit with a new version either from a different organism or with a precisely altered gene sequence. Firstly, RUBISCO is a very sluggish enzyme in its CO_2 fixation reaction and secondly, it also acts as an oxygenase, allowing oxygen to bind at the active site and compete with CO_2. The problem then becomes determining which amino acids could be changed in order to improve enzyme activity. Such experiments have been carried out extensively in photosynthetic bacteria but, with transplastomic technology now available, altering large subunit sequences in transplastomic plants is now a real option. Unfortunately, a lack of understanding of which amino acids or permutation of amino acids to change, and the fact that evolution has probably perfected RUBISCO activity to a maximum, mean that such a goal may not be viable. Alternatively, looking at the enzyme's relative binding of oxygen versus carbon dioxide could, in theory, improve the balance towards carboxylation and reduce the wasteful process of photorespiration. Such a goal is aided by the fact that the CO_2:O_2 specificity in the RUBISCO enzyme from sunflower (*Helianthus annus*) is 10% greater than in the tobacco enzyme. Thus, generation of a transplastomic in tobacco, in which the tobacco large subunit was replaced with the sunflower version was achieved and yielded functional RUBISCO enzymes with eight tobacco small subunits and eight sunflower large subunits. Unfortunately, the resulting hybrid RUBISCO enzyme did not show the anticipated biochemical properties and such experiments starkly reveal the unpredictability of such approaches. Undoubtedly, further efforts will be made in this area in the future but outcomes of such experiments may not be predictable.

In theory, changes could be made to any of the plastid-encoded components of the photosynthetic transport system on the thylakoid membrane, with a view to trying to improve the efficiency of different aspects of the process. It is clear, however, that a very detailed understanding of precisely how amino acids function within each protein is needed in order to identify possible sites that could be changed to improve efficiency, an understanding that at present is probably out of reach.

Manipulating biochemical metabolic pathways in plastids would appear to be a somewhat easier goal than that of fiddling about with photosynthesis. Carotenoids are an important group of molecules, which have many health benefits when consumed. Vitamin A is synthesised in the human body from provitamin A, which is β-carotene. In addition, several other carotenoids, found mostly as accessory pigments in the chlorophyll antennae complexes associated with the two photosystems on the thylakoid membrane, are important antioxidants in the human diet, including zeaxanthin, lutein and lycopene. Since carotenoid biosynthesis takes place entirely in the plastid (see Chapter 7), efforts to manipulate levels of carotenoids by manipulating biosynthetic enzymes in the pathway could either involve targeting increased amounts of relevant enzymes to the plastid from a nuclear transgene or transforming the plastid genome directly. A classic tale of plant genetic engineering was the enhanced production of β-carotene in the endosperm tissue of rice seeds by nuclear transformation with three enzymes in the pathway and turning on the pathway in endosperm tissue where it normally is not expressed. This resulted in the so-called golden rice, which has the potential to overcome vitamin A deficiency in countries where rice is a major part of the diet. In another project, the elevation in levels of β-carotene in tomato fruit was achieved by expressing the bacterial enzyme lycopene β-cyclase on the plastid genome in transplastomic tomato plants. In these fruit, the major coloured carotenoid pigment lycopene, which accumulates in the chromoplasts, was converted to β-carotene, resulting in a four-fold increase in β-carotene levels. Such a success suggests that manipulation of important dietary components in fruits and seeds by both nuclear and transplastomic technology may well become commonplace in the future.

Other targets for plastid genetic engineering are the synthesis of fatty acids and amino acids, which take place within the plastid (see Chapter 7). An important enzyme in fatty acid synthesis is the enzyme acetyl CoA carboxylase that is normally encoded in the plastid genome. Introduction of a second copy of this gene into the plastid genome with a stronger promoter can result in increased fatty acid content within the leaf tissue.

Plastid transformation and molecular farming

There is great potential in using transgenic plants to synthesise important molecules that are used in the pharmaceutical industry. This area of plant biotechnology has been termed molecular farming and offers the prospect of producing important biopharmaceuticals within plant tissues with relatively low costs of production and delivery. Using transplastomic technology to synthesise foreign proteins, which could be antibodies, antigens or antimicrobial molecules, would appear to be feasible especially since plastids have been shown to be good at accumulating foreign proteins within them from a transplastomic gene.

The first antigen to be synthesised in transplastomic chloroplasts was against the bacterium *Clostridium tetani*, which causes tetanus. Expression of a gene encoding subunit C of the toxin produced by the bacteria on the tobacco plastid genome resulted in significant accumulation of this antigenic protein, with levels up to 25% of total soluble leaf protein, especially when gene sequences optimised for chloroplast codon usage were used. Most importantly, the molecule produced in the plastid was shown to be biologically active in mice and conferred a degree of immunisation against the *Clostridium* toxin.

Transplastomic approaches have also been successful in producing a vaccine against anthrax. The *pagA* gene, which encodes a component of the protein mixture made by the bacteria, *Bacillus antharcis*, is not toxic itself, but is strongly immunogenic and thereby makes a good candidate for development of a vaccine against anthrax. When a transplastomic tobacco plant was made, with the *pagA* gene encoded on the plastid genome, the plastid produced stable antigen protein, which biologically works as well as the same protein produced in bacteria and confers immunity in mice when exposed to the anthrax toxin.

Both of these trial studies show the great potential that exists in using transplastomic technology for the expression of pharmaceutically important molecules. There is one problem, however, as mentioned earlier, in that plastids are not supposed to contain systems for the complex post-translational modification of proteins, which occurs extensively in proteins synthesised in the cytosol. In many cases, post-translational modification, such as glycosylation or lipidation: the adding of sugar and lipid moieties to the protein in question, is critical for the protein to develop its functionality, and transplastomic expression of proteins requiring such modifications may result in a biologically inactive protein. However, in one example of transplastomic vaccine production, lipidation of the foreign protein was shown to occur. The surface lipoprotein A, termed OspA from the bacterium *Borrelia burgdorferi*, is a vaccine against Lyme disease and, when the gene for OspA is expressed on the plastid genome, the resultant protein is lipidated somehow and shows good biological activity as a vaccine, inducing protective antibodies in mice. Thus, there may well be post-translational modification systems present in plastids, which are as yet unknown in detail.

Another approach in transgenic vaccine production in plants has been the idea of oral vaccination, in that a person eats a transgenic plant or fruit thereof, which synthesises transgenic vaccine and thereby the person becomes inoculated against the chosen infection. To optimise such a reaction, adjuvants are required along with the vaccine to enhance the immune response. One molecule that works well in this role is the non-toxic B subunit of the *Cholera* toxin, CTB. It can be expressed trans-plastomically to high levels and can be successfully fused with green fluorescent protein in a chimeric protein construct. In the future, its fusion with an edible vaccine, both expressed together in a gene construct on the plastid genome, may enhance the efficacy of edible vaccines produced transplastomically.

As we have seen in these examples, transplastomic technology has great potential for many aspects of plant biotechnology and molecular farming, although there remain several major problems to its implementation. Not least is the information required to maximise expression from the plastid genome and determination of which type of promoter to use to drive the plastid transgene. Also, knowledge is needed of where is the best site for integration in the plastid genome and what level of post-translational modification might be required. Furthermore, much of the current research work has focused on expression in leaf chloroplasts, whereas extraction of biologically active pharmaceuticals would most probably happen in seeds or fruits. This requires a thorough understanding of the control systems that moderate plastid gene expression in types of plastids, other than chloroplasts, such as amyloplasts, chromoplasts and leuco-plasts, information that at present is lacking. Also, managing expression patterns in a tissue-specific way would be desirable, such that the trans-plastomic gene is only expressed in seeds, for instance, and not in leaves.

It is clear, however, that genetic manipulation of the plastid in these ways will become an important part of plant biotechnology in the future.

Further reading and resources

There is a plethora of research journal articles and research papers available in the scientific community in which the reader will find significantly greater detail about plastid biology than this book provides. Good starting points for further exploration into plastid biology are the various papers listed in the figure legends in this book.

There are many excellent review journals with articles covering many aspects of plastid biology, which include *Trends in Plant Science*, *Trends in Cell Biology*, *Current Opinion in Plant Biology*, *Current Biology* and *Annual Reviews in Plant Biology*. In addition, there are many primary research journals that publish research papers specifically in plant biology, including research on plastids. The main ones are *The Plant Cell*, *The Plant Journal*, *Plant Physiology*, *Journal of Experimental Botany*, *Annals of Botany*, *Plant and Cell Physiology*, *Planta*, *Physiolgia Plantarum* as well as many others. A list is available at www.e-journals.org/botany/#P.

Several multi-authored books in recent times have reviewed many aspects of plastid biology and are an excellent source of more detailed information about plastids.

Cell and Molecular Biology of Plastids (2007) (ed. Bock R). *Topics in Current Genetics* 19. Berlin, Heidelberg, Springer-Verlag. ISBN 978-3-540-75375-9.

The Structure and Function of Plastids (2006) (eds. Wise RR, Hoober JK). *Advances in Photosynthesis and Respiration* 23. Berlin, Heidelberg, Springer-Verlag. ISBN 10-4020-4060-1.

Plastids (2005) (ed. Møller S). *Annual Plant Reviews* 13. Oxford: Blackwells. ISBN 1-4051-1882-2.

Chloroplast Biogenesis: From Proplastid to Gerontoplast. Biswal UC, Raval MK (2003). Kluwer Academic Publishers. ISBN-10: 1402016026.

Another excellent resource is *Plant Biology* on DVD, which is composed of a series of chapters on different aspects of plant cellular structure, including an outstanding chapter on plastids, which contains a wide array of stunning images and movies of many of the processes and structures described in this book. The project has been put together by Professor Brian Gunning, at the Australian National University and is ideal for anyone interested in the structural aspects of plastids and plant cell biology in general (http://www.plantcellbiologyonDVD.com/).

A series of seminar-style audio-visual presentations are available on CD, marketed by Henry Stewart Talks, called 'The chloroplast: the organelle that sustains us'. This CD contains 18 narrated talks with visual slides and covers a whole range of molecular and biochemical aspects of plastid biology, and is ideal for anyone wanting narrated explanation of a wide range of aspects of plastid biology (www.hstalks.com).

Internet resources are also available for plastid biologists, especially for molecular genetic resources and databases; http://www.plastid.msu.edu/ is a good starting point.

Those interested more specifically in the photosynthetic aspects of chloroplasts should look at http://www.life.uiuc.edu/govindjee/photoweb/, which covers all aspects of photosynthesis.

Index

14-3-3 proteins 84, 137

aadA gene 182
abscisic acid (ABA) 151, 152
acetyl CoA 147, 148
acetyl CoA carboxylase 50, 147, 148, 192
actin microfilaments 162, 167–168
acyl-tRNA synthetase 59, 97
adenine 37
adenylate transporter 102
ADP glucose 131
Agrobacterium tumifaciens 179–182
airspaces 169
alanine 139
ALB3 92, 118
aleurone 133
algae 1, 5, 7, 35, 37, 40, 51
algae – green 2, 6, 7, 42, 51
algae – red 6, 8, 51
amino acid biosynthesis 137–144
amino acids 81
ammonium ions (NH$_4^+$) 105, 137, 138
amylase 132, 133–135
amylopectin 20, 21, 133
amyloplast 19–23, 27, 130, 131, 133,
    135, 156, 171
amyloplast differerentiation 176
amylose 20, 133
angiosperms 1, 2, 9, 12, 23, 29, 50, 185
antenna complex 23, 63, 66, 72, 117,
    122, 129
anthocyanin 23
anthrax 193
antibiotic resistance genes 186
Apicomplexa 51
apicoplast 51

aplastidic 11
Arabidopsis 8, 11, 15, 22, 26, 28
arbuscules 175
ARC3 156–157
ARC5 157
ARC6 156, 161
asparagine 143
aspartate 139, 140
ATP 69, 70–72
ATP synthase 43, 54, 65, 70–72, 73
auxin 20, 23, 176

Bacillus antharcis 193
Bacillus thuringiensis 188
bacterial cell division 155
bell pepper 23, 24, 171
binary fission 6, 11, 154
biolistics 179–182, 185
biparental inheritance 12
blue light 120, 167
broken circles 32
bryophytes 9
Bt toxin 188–189
bundle sheath cells 172, 173

C$_3$ photosynthesis 75, 172
C$_4$ photosynthesis 16, 80, 101, 172–174
C$_4$ photosynthesis – single cell 174
cacti 26–27
calcium ions (Ca^{2+}) 104
calvin cycle 75, 76, 130, 132
capsanthin 24
capsorubin 24
carbon dioxide (CO$_2$) 61, 76, 172
carbon dioxide (CO$_2$) fixation 75–80
carbonic anhydrase 97–98

197

carotene 23
carotene-α 24
carotene-β 23, 24, 152, 192
carotenoid biosynthesis 150–152
carotenoids 23, 24, 27, 126, 129, 192
carrot 24, 41, 183, 189
CASTOR 175
catenanes 33
cell suspension cultures 183
cellulose 61
chaperone 77, 84, 85, 86
Chenopodiaceae 174
chloramphenicol 126
Chlorella 42, 51
chlorophyll 1, 3, 23, 28, 29, 62–67, 116, 126
chlorophyll a 62, 63, 114, 115, 116, 118
chlorophyll b 62, 63, 114, 115, 116, 118
chlorophyll synthesis 53, 113–117, 120
chlorophyll synthesis – Light 114, 115
chlorophyllide 29
chlorophytes 1, 2
chloroplast 1, 2, 13–18, 36, 107, 170
chloroplast accumulation response 165
chloroplast avoidance response 165
chloroplast development 29, 88,
    108, 129
chloroplast development – light 107,
    117–120
chloroplast differentiation 107
chloroplast population 155
cholera 194
chorismate 142, 143
chromoplast 23–25, 29, 130, 150,
    159, 164, 171
CHUP1 protein 167
citrus fruit 171
Clostridium tetani 193
columella cells 22, 133, 176
copper ions (Cu^{2+}) 104
copy number 37
co-translational import 59, 124
cotton 183
cotyledons 19, 107
cre-lox recombinase 186
cryptochrome 120–121
cryptophytes 8
Cuscuta 50
cyanobacteria 3, 4, 5, 8, 40, 95
cysteine 144–147
cytidine 37
cytochrome b_{559} 122

cytochrome b_6f complex 43, 65, 68,
    69, 72, 123
cytochrome f 123
cytokinesis 11
cytokinin 20, 171
cytoskeleton 12, 161

D1 protein 59, 65, 67–68, 120, 122, 123, 124
D2 protein 65, 67–68, 120, 122
DAPI (4′, 6-diamidino-2-phenyl indole) 35
diacylglycerol 112
dicistronic 54
dinoflagellate 8, 51
disulphide bond 145
DNA 164
DNA binding proteins 35
DNA gyrase 35
DNA polymerase 35
DNA replication 33
drought stress 189
dual targeting 83, 97

editosome 56
egg cell 12
elaioplast 25–27
electron transfer 67–70
electrostatic interaction 110
elongation factor 59
embryophytes 1
endoplasmic reticulum 83, 89, 112,
    150, 185
endosperm 19, 21, 27, 133
endosymbiosis 3, 5, 6, 9, 31
endosymbiosis – secondary 5, 7, 8
envelope lumen 81
envelope membrane – inner 34, 81, 100, 112,
    113, 115
envelope membrane – outer 16, 81, 85, 99
envelope proteins – outer 99
environmental containment
    of transgenes 185
epidermis 15
Epifagus 50
epigenetic effects 185
epinasty of synthesis (CES) 123
EPSP synthase 142, 143, 187–188
ethidium bromide 32, 33
ethylene 147
etioplast 29–30, 107, 108, 115, 171
eukaryotic Cell 3
evolution 2, 7

fatty acid biosynthesis 50, 148
fatty acid synthetase (FAS) 147–150
ferns 1
ferredoxin 69–70, 135, 138
ferretin 103
ferrodoxin- NAD(P)+ oxidoreductase
  (FNR) 87
fibrillin 25, 26–27
fibrils 24, 25
fleshy fruit 23, 24, 171
fluorescence 67
fructose 2–6 bisphosphate 131
fructose 6-phosphate 76, 131, 132
FtsZ 155–157, 158, 160

galactolipids 112
gene silencing 185
genetic hybrid 43
genome copy number 183
germination 107
germination – epigeal 107
germination – hypogeal 107
gerontoplast 28–29, 171
gibberellin 133, 171
glaucophytes 6, 7
GLK genes 172
glucose 101, 132
glucose 6-phosphate 133, 134, 136
glutamate 114, 138, 139
glutamine:oxoglutarateaminotransferase
  (Fd-GOGAT) 138, 139
glutamine synthetase (GS) 101, 138, 139
glutamyl tRNA 53
glutathione 121, 146–147
glycine 143
glycine betaine 189–190
glycerolipids 149–150
glycosylation 98, 185, 193
glyphosate 142, 187
Golgi bodies 98, 185
grana 12, 29, 111
granal lamellae 109
granal stack 17, 72, 109, 110
gravitropism 23
gravity sensing 22
green fluorescent protein (GFP) 12, 28, 34,
  35, 98, 161, 163, 168, 179
GTP 86
GTPase 112
guanidine 37
guard cell 15

gun mutants 115, 126
GUN1 127, 128
gymnosperms 1, 2, 9, 13, 115

haem 50, 114, 115
haptophytes 8
heavy metals 190
herbicide resistance 186
heterokonts 8
histidine 142
homologous recombination 37,
  181, 183
homoplasmy 182, 185
hornworts 1
hydrogen peroxide 129
hypocotyl cells 163, 164

insect pest resistance 188
intron 54–56
inverted repeat 37–40, 41, 42, 50, 58
ion transport 99
iron ions (Fe^{2+}) 103
iron–sulphur complex 65, 69, 146, 147
isoleucine 139–141
isoprenoid 26–27

JAC1 167

Kranz anatomy 172

large single copy (LSC) 37, 41
leaf hair cells 162, 163
leaf primordia 107
Leguminosae 40
leucine 141
leucoplast 27–28, 130, 159, 162, 171, 174
LHCI 69
LHCII 74–75, 117
LHCII kinase 75
light capture 62–67
light harvesting chlorophyll binding
  proteins (LHC) 63, 65, 66, 83, 92, 118, 126
lignin 143
linocomycin 126
lipid 26–27, 81, 95, 147–150
lipid synthesis 111–112, 122, 144, 173
lipidation 193
liverwort 1, 2, 37
Lolium temulentum 29
LOV domain 167
lutein 63, 152, 192

lycopene 23, 24, 152, 192
lycophytes 2
Lymes disease 2
Lynne Margulis 3
lysine 139–141

magnesium ions (Mg^{2+}) 103, 116
maize (*Zea mays*) 16, 172
maltose 102, 132, 135
manganese ions (Mn^{2+}) 67, 104
marigold (*Tagetes*) 25
maternal inheritance 12
mechanosensing proteins 165
mercury 190
meristem 10–11, 153
mesophyll cell 13–15, 16, 34, 36, 154, 156,
   166, 176–177
metabolic engineering 190
metabolite transport 99
methionine 139–141, 145, 146, 147
Mg-protoporphyrin IX 114, 126, 128
Min genes 42, 51, 155
mitochondrion 71, 80, 83, 87, 97,
   144, 165
molecular farming 193–194
monilophytes 2, 9
monocistronic 54, 55
monoterpenes 150
mosses 1, 2
multigenomic 33
mutation rate 37
myosin motor protein 162, 167

NADPH 65, 69–70, 135–137
naladixic acid 126
neoxanthin 63, 118
nitrate ions (NO_3^-) 137–138
nitrate reductase 137, 138
nitrite ions (NO_2^-) 105, 138
nitrite reductase 137, 138
nitrogen assimilation 99, 137–144
non-green plastid 27, 130, 134, 136,
   150, 162, 164
norflurazon 126
nuclear genome 42, 179
nuclear membrane 164
nuclear transcription factors 172
nuclear transformation 183
nucleoid 34, 36, 89, 108, 185
nucleolus 11
nucleus 11, 12, 14, 83, 163, 164

operon 185, 188
osmoprotectant 189–190
oxidative pentose phosphate pathway
   (OPPP) 135–137
oxygen 68
oxygen evolving complex (OEC) 68, 90,
   104, 172
oxygenase 172

P680 67
P700 67, 69
PDV1 157
PDV2 157
pea (*Pisum sativum*) 19
pelargonium 50
pentatricopeptide repeat protein (PPR) 57
peptidoglycan 6
pericarp cells 162
peri-nuclear array 164
peroxisome 67, 143, 165
petal 23, 27, 171
phagotrophy 3, 5
phenylalanine 142, 143
pheophytin 67, 68
phosphate translocator 100–101
phosphinothricin 188
phosphenolpyruvate 101
phosphoenolpyruvate (PEP) carboxylase
   101, 172
phosphoglycolate 80
photoinhibition 123, 125
photolysis 67–68
photomorphogenesis 118
photoreceptor 118
photorespiration 80, 99, 101, 139
photosynthesis 1, 7, 50, 191, 61–80
photosynthetic bacteria 61, 115
photosynthetic proteins 43
photosystem I 43, 62–67, 69, 72, 73, 90
photosystem II 43, 62–67, 72–74, 90, 122,
   124, 172
photosystem II repair cycle 74, 123–125
phototropic growth 167
phototropin 167
phycobilosomes 4
phycocyanin 3
phycoerythrin 3
phylloquinone 150
phytochrome 108, 118–120
phytoene 23
phytoene desaturase 25, 152

phytoene synthase 25, 152
phytol 28
phytoremediation 190
pine 50
plant pathogens 190
plants – flowering 1
plants – land 1, 2, 6, 37
plants – lower 1, 115
plants – parasitic 50
Plasmodium 51
plastid biotechnology 178–194
plastid compartment size 178
plastid differentiation 153, 169, 170
plastid dividing ring (PD) 155–157, 160
plastid division 36, 51, 107, 154, 156, 160
plastid DNA 4, 32–38, 126, 180, 185
plastid envelope 10, 21, 88, 90, 93, 99, 112, 132, 145, 155, 156–157, 160, 168
plastid envelope binding protein (PEND) 34, 35
plastid gene expression 120–122, 126–127, 128
plastid genome 3, 31, 34, 38–40, 41, 42, 44–49, 179
plastid genome sequence 37
plastid import 81–105
plastid inheritance 12
plastid metabolism 130–152
plastid morphology 159–165
plastid movement 165–169
plastid nucleoids 34–37
plastid proteome 95–98
plastid transcription kinase (PTK) 121
plastid transformation 178–194
plastocyanin 65, 68, 69, 89, 90
plastoglobuli 16, 18, 26, 27, 28, 29
plastoquinone 27, 65, 68–69, 74, 121, 127, 150
plastoskeleton 161
pollen 12, 26–27, 185
POLLUX 175
polycistronic 53–54, 55
polyploid 34
polyprotein 94
poplar 183
porphyrin 116
potato (*Solanum tuberosum*) 21, 171, 183
porin 100
post-translational modification 185
prokaryote 3
prolamellar body 29–30, 115
proline 143

proplastid 10–13, 19, 36, 106, 107, 153, 170, 185
protein degradation 43
protein turnover 122
proteinoplast 27, 28
proteome 35, 41, 171, 173, 178
protochlorophyllide 29, 115
protochlorophyllide oxidoreductase (POR) 29, 114, 115, 120
pulse-field gel electrophoresis (PFGE) 32
pumpkin 171

Q_A 68
Q_B 68

rDNA 54
reaction centre 66
reactive oxygen species 127
recombinant proteins 98
recombinase – RecA 181
recombination 33, 37, 40, 180
red light 118
redox 87
redox control 121–122, 135
reductive pentose phosphate pathway (RPPP) 130
resonance 59
retinal 152
ribosomal – plastid specific proteins 58
ribosomes 4, 6, 10, 18, 37, 43, 57–60, 85, 164
ribosomes – membrane bound 59
ribulose bisphosphate (RuBP) 76, 78, 79
RNA 164
RNA binding proteins 52, 55, 57
RNA editing 56
RNA polymerase 35, 51, 53, 120, 121
RNA polymerase – NEP 52–53
RNA polymerase – PEP 50, 52
RNA processing 43, 53–57
RNA processing proteins 52
RNA splicing 54–56
RNA stability 56
RNA transplicing 56
root 27, 137, 163
root meristem 12
root plastid 27, 28, 130, 164, 174–176
rRNA 40, 43
RUBISCO (ribulose-1,5 bisphosphate carboxylase/oxygenase) 18, 28, 43, 75, 76, 126, 172, 191

RUBISCO activase 78
RUBISCO large subunit (LSU) 43, 54,
    77–78
RUBISCO small subunit (SSU) 43,
    77–78, 83

salinity stress 189
sec pathway 90–91, 93, 124
secondary pigments 63
seed dispersal 23
segregation 11–12
selectable marker gene 180, 182
senescence 28
serine 143
shikimate pathway 142, 143
shoot apical meristem 10, 11, 107
shuttle vectors 186
sigma factors 52, 120, 121
signalling 107, 131
signalling – plastid to nuclear 108,
    125–129, 165
signalling – redox 68, 127, 128
signalling – anterograde 125
signalling – retrograde 125, 126–127, 128
small single copy (SSC) 37, 41
solanin 171
somatic embryos 183
source-sink 135
soybean 32, 183
spectinomycin 180, 182
starch 18, 20–22, 76, 79, 102, 132
starch branching enzyme 132, 133
starch breakdown 133
starch granule 19, 20–22, 29
starch synthase 10, 83, 131, 132
starch synthesis 130–135
state II – state I transition 75, 121
statolith 22, 23, 176
stomata 15, 167
stop-transfer domain 89
stroma 16, 17, 18, 93
stromal lamellae 17, 18, 109, 110, 111
stromal processing peptidase (SPP)
    85, 87, 89, 118, 188, 94
stromal signal recognition particle (SRP)
    92, 93, 118
stromule 28, 99, 162–165, 175
sucrose 76, 79, 102, 132
sucrose synthesis 130–135
sugar beet (*Beta vulgaris*) 131
sugar cane (*Saccharum officinarum*) 131, 172

sulphate ions ($SO_4{}^{2-}$) 105, 144, 145
sulphate transporter 144
sulphide ions (S^{2-}) 144, 146
sulphite ions ($SO_3{}^{2-}$) 144, 145
sulphite reductase 35
sulpholipids 144, 145
sulphur dioxide 144
sulphur metabolism 99, 144–147
sunflower (*Helianthus annuus*) 191
superoxide dismutase 104
symbiotic fungi 175
Synechocystis 4, 40, 95, 96

tapetal cells 26–27
tat pathway 26–27, 90, 91–92, 93
T-DNA 179
temperature stress 189
tetracistronic 54
tetrapyrrole 113, 173
thioredoxin 70, 71, 135, 145
threonine 139–141
thylakoid lumen 18, 68–69, 70, 86,
    89, 90, 93
thylakoid membrane 4, 5, 10, 11,
    16–18, 24, 25, 26–27, 28, 29, 30,
    57, 58, 59, 62, 65, 72–75, 89, 93,
    109–113, 172
thylakoid membrane synthesis 113
thylakoid membrane targeting sequence 92
thylakoid processing peptidase (TPP) 90, 91,
    93, 94
thylakoid protein complex assembly
    122–125
thymidine 37
Tic complex 84, 85, 118
Tic Toc import systems 95, 179
Ti-plasmid 179
tobacco (*Nicotiana Tabacum*) 12, 34,
    164, 182
Toc Complex 84, 85, 118
Toc Isoforms 88
TOC75 95, 96
tocopherol 27, 150
tomato 23, 24, 152, 162, 171, 183
tonoplast 15, 155
topoisomerase 35
toxoplasma 51
transcription 43, 51–53
transcription factor 120
transcription factor – ABI4 127
transgene 179

transit peptide 82–83, 90, 94, 96, 97
transit peptide subfragment degrading
    enzyme 87
translation 43, 57–60, 126
translocon 84–87
transport proteins 99
trehalose 189
triose phosphate 79, 131, 132
triose phosphate transporter (TPT)
    100–101, 131, 132
tRNA 43, 54, 59
tryptophan 142, 143
tubulin 155
tyrosine 142, 143

UDP-glucose 131

vaccine production 193–194
vacuole 13, 15, 23, 155
valine 141
van der Waals' force 110
vesicles 112–113, 117, 118, 162, 164
violoxanthin 23, 63, 118
vitamin A 24, 192
vitamin K 69

wheat (*Triticum aestivum*) 14

xanthophyll 152

ycf genes 50

zeaxanthin 23, 192

Printed in the United States
by Baker & Taylor Publisher Services